CW00642396

CONSTRUCTION ROBOTS

Elementary Technologies and Single-Task Construction Robots

Single-task construction robots (STCRs) are robots developed for use on the construction site. After the first experiments in large-scale prefabrication were successfully conducted in Japan, and the first products proved successful on the market, the main contractor, Shimizu (1975, Tokyo), set up a research group for on-site construction robots. The focus initially was set on simple systems able to execute single, specific construction tasks in a repetitive manner. Today STCRs are a worldwide research and development theme, and the tasks and application fields of STCRs expand continuously. Whereas the first approaches done in Japan built on relatively simple manipulators and mobile platforms used to distribute concrete, finish floors, install wall panels and move material, recently new forms of STCRs have emerged building on aerial approaches, additive manufacturing technologies, exoskeletons, swarm robotic approaches, self-assembling building structures and humanoid robot technology. This volume features 200 STCRs classified into 24 categories.

Thomas Bock is a professor of building realization and robotics at Technische Universität München (TUM). His research for the past 35 years has focused on automation and robotics in building construction, from the planning, prefabrication, on-site production and utilization phases to the reorganization and deconstruction of a building. His educational and professional experience is mostly from Europe, United States and Japan. He is a member of several boards of directors of international associations and of several international academies in Europe, the Americas and Asia. He consulted several international ministries and evaluates research projects for various international funding institutions. He holds honorary doctor and professor degrees, as well as fellowships and visiting professorships. Professor Bock serves on several editorial boards, heads various working commission and groups of international research organizations, and has authored and co-authored Cambridge University Handbook Series on Construction Robotics and more than 400 articles.

Thomas Linner is a research associate in building realization and robotics at Technische Universität München (TUM). During the last few years, he has supervised some major research projects, with a focus on the deployment of advanced technology in the building sector. He is a specialist in the area of automated production of building products as well as in the enhancement of the performance of building products by advanced technology. He completed his dissertation in the field of construction automation, focussing on automated/robotic on-site factories. Increasingly, the generation of innovation strategies, business models, value systems and innovative manufacturing organization methods complementary with advanced technology in construction is becoming the central issue in his research. Dr. Linner has been an invited speaker at universities such as the University of Tokyo and Cambridge University. He has received several prizes and grants, including a Japanese Center of Excellence Grant for research in Japan.

CAMBRIDGE HANDBOOKS ON CONSTRUCTION ROBOTICS

The Cambridge Handbooks on Construction Robotics series focuses on the implementation of automation and robot technology to renew the construction industry and arrest its declining productivity. The series is intended to give professionals, researchers, lecturers, and students basic conceptual and technical skills and implementation strategies to manage, research, or teach the implementation of advanced automation and robot technology–based processes and technologies in construction. Currently, the implementation of modern developments in product structures (modularity and design for manufacturing), organizational strategies (just in time, just in sequence, and pulling production), and informational aspects (computer-aided design/manufacturing or computer-integrated manufacturing) are lagging because of the lack of modern integrated machine technology in construction. The Cambridge Handbooks on Construction Robotics books discuss progress in robot systems theory and demonstrate their integration using real systematic applications and projections for off-site as well as on-site building production.

Robot-Oriented Design: Design and Management Tools for the Deployment of Automation and Robotics in Construction ISBN 9781107076389

Robotic Industrialization: Automation and Robotic Technologies for Customized Component, Module, and Building Prefabrication ISBN 9781107076396

Site Automation: Automated/Robotic On-Site Factories ISBN 9781107075979

Construction Robotics: Elementary Technologies and Single Task Construction Robotics ISBN 9781107075993

Ambient Robotics: Automation and Robotic Technologies for Maintenance, Assistance, and Service ISBN 9781107075986

Construction Robots

ELEMENTARY TECHNOLOGIES AND SINGLE-TASK CONSTRUCTION ROBOTS

Thomas Bock
Technische Universität München

Thomas Linner
Technische Universität München

CAMBRIDGE
UNIVERSITY PRESS

One Liberty Plaza, 20th Floor, New York, NY 10006, USA

Cambridge University Press is part of the University of Cambridge.

It furthers the University's mission by disseminating knowledge in the pursuit of education, learning, and research at the highest international levels of excellence.

www.cambridge.org
Information on this title: www.cambridge.org/9781107075993

First published 2016

Printed in the United States of America by Sheridan Books, Inc.

A catalog record for this publication is available from the British Library.

Library of Congress Cataloging in Publication Data
Names: Bock, Thomas, 1957– | Linner, Thomas, 1979–
Title: Construction robots / Thomas Bock, Technische Universität München, Thomas Linner, Technische Universität München.
Description: Cambridge : Cambridge University Press, 2016. | Includes bibliographical references and index.
Identifiers: LCCN 2016021809 | ISBN 9781107075993 (hardback : alk. paper)
Subjects: LCSH: Robots, Industrial. | Robots – Design and construction.
Classification: LCC TS191.8 .B6285 2016 | DDC 629.8/92–dc23
LC record available at https://lccn.loc.gov/2016021809

ISBN 978-1-107-07599-3 Hardback

Contents

Acknowledgements

Construction automation gained momentum in the 1970s and 1980s in Japan, where the foundations for real-world application of automation in off-site building manufacturing, single-task construction robots, and automated construction sites were laid. This book series carries on a research direction and technological development established within this "environment" in the 1980s under the name *Robot-Oriented Design,* which was a focal point of the doctoral thesis of Thomas Bock at the University of Tokyo in 1989. In the context of this doctoral thesis many personal and professional relationships with inventors, researchers, and developers in the scientific and professional fields related to the construction automation field were built up. The doctoral thesis that was written by Thomas Linner (*Automated and Robotic Construction: Integrated Automated Construction Sites*) in 2013 took those approaches further and expanded the documentation of concepts and projects. Both of these form the backbone of the knowledge presented in this book series.

The authors first wish to express their deepest gratitude to Prof. Dr. Yositika Uchida, Prof. Dr. Y. Hasegawa, and Prof. Dr. Umetani (TIT); Dr. Tetsuji Yoshida, Dr. Junichi Maeda, Dr. Yamazaki, Dr. Matsumoto, Mr. Abe, and Dr. Ueno (Shimizu); Prof. Yashiro, Prof. Matsumura, Prof. Sakamura, Prof. Arai, Prof. Funakubo, Prof. Hatamura, Prof. Inoue, Prof. Tachi, Prof. Sato, Prof. Mitsuishi, Prof. Nakao, and Prof. Yoshikawa (UoT); Dr. Oiishi and Mr. Fujimura (MHI); Dr. Muro, Kanaiwa, Miyazaki, Tazawa, Yuasa, Dr. Sekiya, Dr. Hoshino, Mr. Arai, and Mr. Morita (Takenaka); Dr. Arai, Dr. Chae, Mr. Mashimo, and Mr. Mizutani (Kajima); Dr. Shiokawa, Dr. Hamada, Mr. Furuya, Mr. Ikeda, Mr. Wakizaka, Mr. Suzuki, Mr. Kondo, Mr. Odakuda, Mr. Harada, Mr. Imai, Mr. Miki, and Mr. Doyama (Ohbayashi); Dr. Yoshitake, Mr. Sakou, Mr. Takasaki, Mr. Ishiguro, Mr. Morishima, Mr. Kato, and Mr. Arai (Fujita); Mr. Nakamura, Mr. Tanaka, and Mr. Sanno (Goyo/Penta Ocean); Mr. Nagao and Mr. Sone (Maeda); Mr. Ohmori, Mr. Maruyama, and Mr. Fukuzawa (Toda); Dr. Ger Maas (Royal BAM Group); Dr. Espling, Mr. Jonsson, Mr. Fritzon, Mr. Junkers, Mr. Andersson, and Mr. Karlson (Skanska); Mrs. Jansson, Mrs. Johanson, Mr. Salminen, Mr. Apleberger, Mr. Lindström, Mr. Engström, and Mr. Andreasson (NCC); Mr. Hirano (Nishimatsu); Mr. Weckenmann (Weckenmann); Mr. Ott and Mr. Weinmann (Handtech); Mr. Bauer and Mr. Flohr (Cadolto); Mr. Kakiwada, Mr. Ohmae, Mr. Sawa, Mr. Yoshimi, Mr. Shimizu, and Mr. Oki (Hazama Gumi); Mr. Masu, Mr. Oda, and Mr. Shuji (Lixil); Mr. Yanagihara, Mr. Oda, Mr. Kanazashi, and

Mr. Mitsunaga (Tokyu Kensetsu); Mr. Okada (Panasonic); Mr. Nakashima and Mr. Kobayashi (Kawasaki HI); Mr. Tazawa (Ishikawajima HI); Mr. Okaya (IHI Space); Dr. Morikawa and Mr. Shirasaka (Mitsubishi Denki); Mr. Maeda, Mr. Matsumoto, and Mr. Kimura (Hitachi Zosen); Mr. Shiroki and Mr. Yoshimura (Daiwa House); Mr. Hagihara and Mr. Hashimoto (Misawa Homes); Mr. Okubo (Mitsui Homes); Dr. Itoh, Mr. Shibata, Mr. Kasugai, Mr. Kato, and Mr. Komeyama (Toyota Homes); Mr. Hirano, Mr. Kawano, and Mr. Takahashi (Toyota Motors); Dr. K. Ohno (+); Dr. Misawa, Dr. Mori, Nozoe, Tamaki, Okada, and Dr. Fujii (Taisei); Prof. P. Sulzer and Prof. Helmut Weber (+); Prof. Dr. Fazlur Khan (+); Prof. Dr. Myron Goldsmith (+ IIT); Prof. A. Warzawski (+ Technion); Prof. R. Navon, Prof. Y. Rosenfeld, Prof. Dr. S. Isaac, Prof. A. Mita, Prof. A. Watanabe, Prof. Kodama, and Prof. Suematsu (Toyota National College of Technology); Mr. Suketomo, Mr. Yoshida, Mr. Oku, and Mr. Okubo (Komatsu); Dr. Ogawa and Mr. Tanaka (Yasukawa); Dr. Inaba (Fanuc); Mr. Usami (Hitachi RI); Mr. Ito and Mr. Hattori (Mitsubishi RI); Mr. Noda, Mr. Ogawa, and Mr. Yoshimi (Toshiba); Mr. Senba (Hitachi housetec); Mr. Sakurai, Mr. Hada, Mr. Sugiura, Mr. Yoshikawa, Mr. Shimizu, Mr. Tanaka, Mr. Nishiwaki, and Mr. Nomura (Sekisui Heim); Mr. Aoki, Mr. Watanabe, Mr. Kotani, and Mr. Kudo (Sekisui House); Mr. Takamoto, Mr. Mori, Mr. Baba, Mr. Sudoh, Mr. Fujita, and Mr. Hiyama (TMsuk); Prof. Dr. T. Fukuda, Prof. Iguchi, Prof. Ohbayashi, Prof. Tamaki, Mr. Yanai, and Mr. Hata (JRA); Mr. Uchida (JCMA); Dr. Kodama (Construction Ministry Japan); Dr. Sarata (AIST); Mr. Yamamoto (DoKen); Mr. Yamanouchi (Ken-Ken); Mr. Yoshida (ACTEC); Prof. Dr. Wolfgang Bley, Prof. Dr. G. Kühn, and Jean Prouve (+); Dr. Sekiya, Dr. Yusuke Yamazaki, Prof. Dr. Ando, Prof. Dr. Seike, Prof. Dr. J. Skibniewski, Prof. Dr. C. Haas, Dr. R. Wing, Prof. Dr. David Gann, Prof. Dr. Puente, Prof. Dr. P. Coiffet, Prof. Dr. A. Bulgakow, Prof. Dr. Arai, Prof. Kai, Prof. Dr. A. Chauhan, Prof. Dr. Spath, Dr. M. Hagele, Dr. J.L. Salagnac, and Prof. Cai (HIT); Prof. Dr. Szymanski (+); Prof. Dr. F. Haller (+); Prof. Dr. Tamura and Prof. Dr. Han.

The authors wish to express their very sincere gratitude to the Japanese guest Professors Dr. K. Endo and Dr. H. Shimizu for supporting and advising the authors during their stays at the authors' chair in Munich. Furthermore, the authors are very thankful to all the companies in the field of component manufacturing, prefabrication, and on-site automation outlined throughout the book series for the information, analyses, pictures, and details of their cutting-edge systems and projects that they shared and provided.

The authors are indebted to the AUSMIP Consortium and the professors and institutions involved for supporting their research. In particular, the authors extend their thanks to Prof. Dr. S. Matsumura, Prof. Dr. T. Yashiro, Prof. Dr. S. Murakami, Dr. S. Chae, Prof. Dr. S. Kikuchi, Assistant Prof. Dr. K. Shibata, Prof. Dr. S.-W. Kwon, Prof. Dr. M.-Y. Cho, Prof. Dr. D.-H. Hong, Prof. Dr. H.-H. Cho, Prof. A. Deguchi, and Prof. B. Peeters and their students and assistants for supporting the authors with information and organizing and enabling a multitude of important site visits. Furthermore, the authors gratefully acknowledge the International Association for Automation and Robotics in Construction, whose yearly conferences, activities, publications, and network provided a fruitful ground for their research and a motivation to follow their research direction.

The authors are indebted in particular to M. Helal, W. Pan, and K. Iturralde for their great support in the phase of completing the volume. The authors thank

Dr.-Ing. C. Georgoulas, A. Bittner, and J. Güttler for their advice and support. In addition, the authors thank I. Arshad, P. Anderson, and N. Agrafiotis.

The authors are sincerely grateful also to their numerous motivated students within the master course Advanced Construction and Building Technology who have, as part of lab or coursework, contributed to the building of the knowledge presented and who have motivated the authors to complete this book series.

1 Introduction

Single-task construction robots (STCRs) were developed predominantly for use on the construction site. After the first experiments in large-scale industrialized, automated, and robotized prefabrication of system houses were successfully conducted in Japan (see **Volume 2**), and the first products (such as Sekisui Heim's M1 prefabricated system) also proved successful in the market, the main contractor, Shimizu (1975, Tokyo), set up a research group to develop on-site construction robots. The goal was now no longer the shifting of complexity into a structured environment (SE) as in large-scale prefabrication (LSP), but the development and deployment of systems that could be used locally on the construction site to create structures and buildings. The focus initially was on simple systems in the form of STCRs that could execute a single, specific construction task in a repetitive manner. The fact that STCRs were task specific made them, on the one hand, highly flexible (they could be used along with conventional work processes and did not require the whole site to be structured and automated), but also represented a major weakness. The fact that in most cases they were not integrated with upstream and downstream processes and the need for safety measures because of the parallel execution of work tasks by human workers in the area where the robots were operating often counterbalanced productivity gains.

Above all, the setup of the robots on-site (equipment, programming) was time consuming and demanded new skills. The relocation of the systems on-site was in many cases complex and time consuming. Therefore, the evaluation of the first generation of developed and deployed STCRs and the identification of the aforementioned problems led step by step, from 1985 onwards, to the first concepts for integrated automated/robotic on-site factories that integrate STCRs and other elementary technology, such as subsystems, into SEs to be set up on the construction site. For this reason, the development of construction robot technology in general and in particular the concept of structuring on-site environments by means of robot-oriented design (ROD) were pushed forward. The conceptual and technological reorientation towards integrated automated construction sites was initiated by Waseda Construction Robot Group (WASCOR), which brought together researchers from all major Japanese construction and equipment firms. The first in-use phase of STCRs, as well as the conceptual reorientation from 1985 onwards, laid the technological and conceptual basis for the development and deployment of automated/robotic on-site factories in the following years.

The development of STCRs provided the basis and paved the way for the realization of integrated automated/robotic on-site factories from the 1990s onwards, and during the past decade, as a result of the flexibility and potential it provides, actually became once more a dominant research field in construction automation and robotics. The development of STCRs parallel to, or as subsystems of, integrated automated construction sites has continued to the present day, and a multitude of new robots have been developed since then and new categories have even emerged. With robot technology becoming more compact and lightweight, significant advances in human–robot interaction and cooperation taking place, and the capability and usability of software for the planning and execution of tasks growing more mature, STCRs are indeed becoming a more and more feasible solution also in less structured environments. The great advantage of the STCR approach is its flexibility and adaptability. It does not require turning the whole construction site into a manufacturing facility, is nonexclusive (and thus allows, if demanded, a focus on the robotization of only individual performance critical, dangerous, or labor intensive tasks), necessitates only a manageable amount of change in terms of construction procedures compared to conventional construction, and designed in the right way, STCRS can be adapted to a huge variety of different site conditions.

Although the STCR approach emerged in Japan, it is today a worldwide research, development, and application theme and STCR development is brought forward in the incumbent industrial nations (Japan, Europe, USA) as well as in catching up industries (in Korea, China, India, Russia, Poland, etc.). Furthermore, the task fields, approaches, and categories of STCRs broaden continuously. Whereas the first systems in Japan built on relatively simple manipulators and mobile platforms used to distribute concrete, finish floors, install wall panels, and move material, recently new forms of STCRs emerged building on aerial approaches, additive manufacturing technologies, exoskeletons, swarm robotic approaches, self-assembling building structures, and even humanoid robot technology.

1.1 History and Development of the STCR Approach

At the end of the 1970s, Shimizu and other Japanese general contractors conducting large building and infrastructure construction projects observed a huge potential in construction robots. Subsequently, with the beginning of the robotics boom in the early 1980s, in which automation and robot technology in all industries in Japan suddenly spread enormously (see also **Volume 1**), the theme became so relevant for the Japanese construction industry that it eventually led to the Japanese government starting, promoting, and bringing forward STCR technology. The chronic shortage of skilled workers in Japan was another reason. Finally, in 1978, the Japan Industrial Robot Association (JARA), under the guidance of the Ministry of Trade and Industry (MITI), established a commission headed by Professor Yukio Hasegawa for the analysis of such applications and the development of automated systems and robotic technology in construction. Participants of this commission were mostly young and motivated engineers from the major Japanese construction companies, and also included general contractors and machine builders. The commission quickly became a "germ cell" for new concepts, and step-by-step specific research projects and robotic systems were set up and implemented by companies.

Numerous universities followed this trend. Waseda University, for example, founded the legendary WASCOR group, which started developing automated and robotic construction technology using an interdisciplinary, cross-sector approach. Then, in the early 1980s, the coordinated activities of the large, national research institutes followed. In 1983, the Architectural Institute of Japan (AIJ) and its commission, responsible for building materials and construction methods, implemented a group (with 15 participating institutions including companies, associations, universities, and public bodies) for automation and robotics in construction. Shortly afterwards, the Japan Society of Civil Engineers (JSCE) followed, and from 1985 the renowned Building Research Institute of the Japanese Ministry of Construction (BRI) started to work with the Center for Development on systems for robotic assembly (e.g., Solid Material Assembly System [SMAS]). In 1987, the Building Contractors Society (BCS), whose members were once more the major construction companies, started with a systematic assessment of the need and potential of automation and robotic technology, in particular for subcontractors, equipment manufacturers, and construction equipment rental companies.

The reasons for the synergistic activities of government, national research institutes, general contractors, and academic institutions had both political and socioeconomic grounds. For example, the low productivity in the construction industry compared to the manufacturing industry, shortage of skilled labour, aging of construction workers, increasingly poor workmanship, rising work-related diseases, and poor working conditions were controversial topics of discussion for the public. The construction industry, which in Japan has traditionally had a high reputation in society, thus faced strong pressure to improve the working environment and the general image of the construction industry.

Figure 1.1 outlines the timeline of activity of the aforementioned institutions participating in the development of STCRs. All in all, the following institutions were involved in the development and deployment of automated and robotic technology during the 1980s and 1990s (first in the form of STCRs and later in the form of integrated automated/robotic on-site factories) for on-site building construction:

- Japan Robot Association (JARA)
- Ministry of Industry and Trade (MITI)
- Waseda Construction Robot Group (WASCOR)
- Ministry of Construction (MOC)
- Architectural Institute of Japan (AIJ)
- Building Contractors Society (BCS)
- Advanced Construction Technology Center (ACTEC)
- Japan Society of Civil Engineers (JSCE)
- Building Research Institute of the Japanese Construction Ministry (BRI)
- Research institutes of large construction companies (Shimizu, Obayashi, Kajima, Maeda, Goyo, Toda, Taisei, Fujita)
- Manufacturers of automation and robot technology

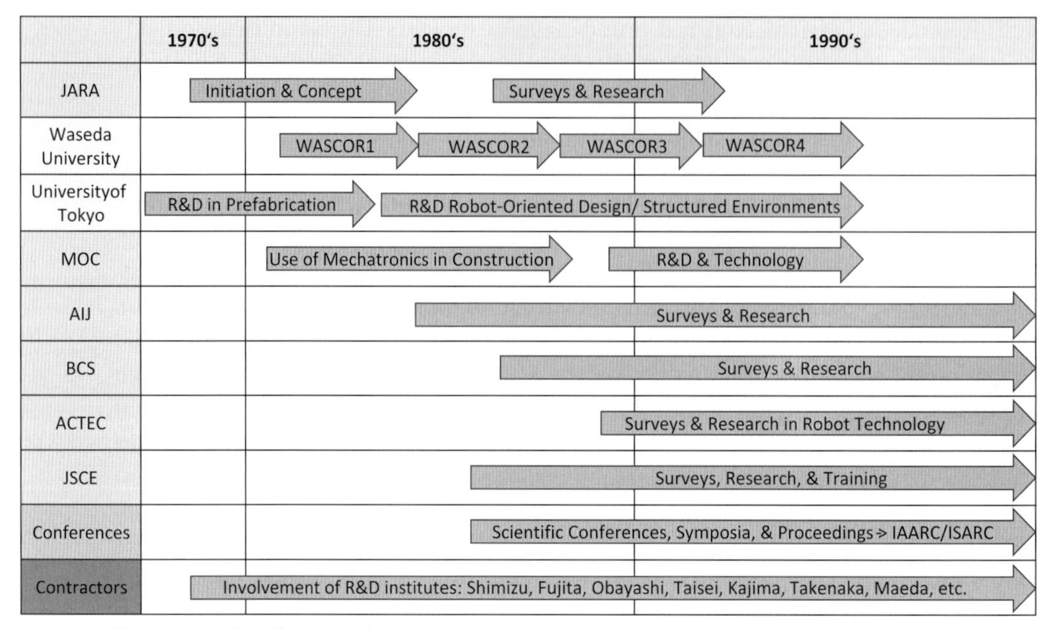

Figure 1.1. Timeline showing activity of institutions participating actively in the development of STCRs (refined and complemented with the authors' information on the basis of Cousineau & Miura 1998 and Hasegawa 1999.)

- Construction/manufacturing equipment suppliers (e.g., Komatsu, Hitachi, Mitsubishi, Kawasaki, Hazama, etc.)
- Universities: Waseda University, The University of Tokyo, etc.

The following research and development (R&D) investment sources contributed to automated and robotic technology (first in the form of STCRs and later as integrated automated sites) for on-site building construction during the 1980s:

- R&D budget of construction companies
- R&D budget of equipment suppliers
- Ministry of Construction (MOC)
- Manufacturers of automation and robot technology

1.2 Strengths and Weaknesses of the STCR Approach

STCRs are systems that support workers on the construction site in executing one specific construction process or task (e.g., digging, concrete levelling, concrete smoothening, brickwork construction, logistics, and painting) or by completely substituting the physical activity of human workers necessary to perform this one process or task. The processes and tasks assisted or executed by STCRs are in most cases relatively profession or craft specific. Furthermore, the processes and tasks for which STCRS were developed have in common that they necessitate a high rate of repetitive subactivities. Further common characteristics are as follows:

1. STCRs are developed predominantly for use on the construction site.

2. STCRs are highly specific, not only to a profession, but even to a task within a specific profession (e.g., different systems for concrete pouring, levelling, and smoothing, which all fall within the realm of the "floor layer" profession)
3. Enhanced productivity compared to conventional labour- and machine-based execution of work tasks:
   a. More m^2/hour than conventional execution (e.g., concrete floor finishing rate – labour based: 100–120 m^2/hour; concrete floor finishing rate robot: 300–800 m^2/hour, according to Cousineau & Miura 1998)
   b. Increased labour productivity
4. Positive impact on quality through precise control of functions and operations (e.g., uniform distribution of paint) and by allowing execution to be recorded or monitored in real time
5. Improvement of working conditions: reduction of dangerous and heavy physical work
6. Various operation modes allowed by most robots: automatic sensor-guided, automatic preprogrammed, remote controlled
7. Positive impact on resource consumption through precise automatic control (e.g., painting robots ensuring that the amount of paint was precisely controlled and that spare paint was collected and reused)
8. In most cases, simple but robust sensor technology: gyroscopes, simple laser measuring systems, touch/pressure sensors, and so forth
9. In many (but not all) cases no more than one operator required to supervise the systems (Systems supervised by two or more persons are inefficient; for further explanation, see **Volume 1.**)

As in any other industry, concepts of modularity developed slowly and step by step over time. Modularity, and thus adaptability to multiple work processes or tasks, was not a characteristic of STCRs in the beginning. This reduced the operational scope of the systems and increased the cost of the robot systems, despite the aforementioned benefits; as such, it not possible to distribute that cost over various work activities (as with conventional multipurpose construction equipment, for example). Some companies introduced modular approaches only in later robot generations, such as by allowing for end-effector change.

The fact that STCRs are task specific makes them on the one hand highly flexible (they can be used along with conventional work processes and it is not necessary for the whole site to be structured and automated), but also presents a major weakness. In most cases they are not integrated with other construction processes, which demands safety measurements and hinders parallel execution of work tasks by human workers in the area where STCRs are operated. As a result, productivity gains are often counterbalanced. Above all, the setup of the robots on-site (equipment transport, task setup/programming) is time consuming and demands skills that extend beyond those of today's construction workers. Furthermore, the relocation of the systems on-site is in many cases complex and time consuming.

Indeed the evaluation of the first generation of developed and deployed STCRs, and the occurrence of the aforementioned problems (which continued technological development today makes it more and more possible to overcome), led step by step from 1985 onwards to the concept of automated/robotic on-site factories. Although

STCR technology was incrementally improved during the 1980s and the 1990s and advanced from the so-called first-generation robots to second- and third-generation robots (outlined in more detail in Cousineau & Miura 1998), the automated/robotic on-site factory approach (given the state of the technological development in robotic and related fields at that time) provided better possibilities to reduce work task spectrum and human labour as well as prestructuring the environment as the basis for higher automation ratios and just in time, just in sequence strategies, such as for factory internal component processing.

However, recent approaches show that major Japanese construction companies today are returning more and more to single-task–like approaches. Obayashi, for example, at present no longer uses its integrated, automated construction site (automated building construction system [ABCS]) as a total system, but applies some of its subsystems as STCRs (e.g., automated logistics systems, welding systems; for further details, see **Volume 4**). By not directly and rigidly connecting those systems, Obayashi gains workshop-like flexibility (in contrast to chain-like organizations to which integrated sites were formerly oriented), which is necessary when constructing buildings such as the Tokyo Skytree, which changes its shape several times from bottom to top. New management approaches, acquired knowledge about the deployment, work process integration of single robotic or automated applications, digital work process management, and increasing usability of the software and interfaces used to set up the tasks to be executed by STCRs today positively influence the integration of such systems in the overall construction process and enhances STCR efficiency compared to the first-generation systems deployed during the 1980s. The development and deployment of STCRs thus is now, when more and more individuality of a product is demanded, more relevant than ever before.

1.3 Analysis and Classification Framework

In the context of producing this volume, STCRs were analysed according to the following framework:

- Background behind development
- Operational capacity
- Technical description
- Control strategy and informational aspects
- Dimensions and workspace
- Description of the robot-supported construction work process and comparison with the conventional work process
- Analysis of the composition and kinematic structures

On the basis of the analysis, 24 categories for task-specific on-site construction robots (STCRs) were defined:

1. Automated site measuring and construction progress monitoring
   a. Mobile robots
   b. Aerial robots
2. Earth and foundation work robots
3. Robotized conventional construction machines

4. Reinforcement production and positioning
5. Automated/robotic 3D concrete structure production on the site
6. Automated/robotic 3D truss/steel structure assembly on the site
7. Bricklaying robots
8. Concrete distribution robots
9. Concrete levelling and compaction robots
10. Concrete finishing robots
11. Site logistics robots
12. Aerial robots for building structure assembly
13. Swarm robotics and self-assembling building structures
14. Robots for positioning of components (crane end-effectors)
15. Steel welding robots
16. Facade installation robots
17. Tile setting and floor finishing robots
18. Facade coating and painting robots
19. Humanoid construction robots
20. Exoskeletons wearable robots and assistive devices
21. Interior finishing robots
22. Fireproof coating robots
23. Service, maintenance, and inspection robots
24. Renovation and recycling robots

The classification follows a work-task oriented approach, as the analysis shows that STCRs do not introduce new organizational settings, but aim at supplementing existing work tasks in conventional and at best slightly altered construction environments. The categorization thus refines and extends existing classifications (Bock 1989; Cousineau & Miura 1998) that followed a similar strategy but did not (as the development continued since then) cover the amount and variety of STCRs considered in this volume. A focus in this volume is on construction robots used on-site to construct buildings. Construction robots used off-site as well to construct, for example, civil infrastructures such as roads, bridges, tunnels, and so forth, are not considered in this volume.

The analysis of STCRs was conducted on the basis of a large picture and information archive; technical data; technical drawings and STCR analysis methodologies introduced by Bock (1989); and technical data, product description brochures, and background information provided by companies and researchers in charge of the development of individual systems. A further but incomplete source with detailed technical drawings was the *Construction Robot System Catalogue* (1999) published by the Japanese Robot Association. Detailed analyses and comparisons of a limited amount of concrete finishing robots, facade painting robots, facade inspection robots, and interior finishing robots by Cousineau and Miura (1998) completed general and technical information of the analysis of systems in those categories. Helpful for the identification of robots developed before the year 2000 was also the catalogue *Robots and Automated Machines in Construction* published by the board of directors of International Association for Automation and Robotics in Construction (1998). Helpful for the identification of robots developed *after* the year 2000 were the *Proceedings of the International Symposium of Automation and Robotics in*

Construction based on a conference at which the "who's who" of specialists involved in the development of construction automation systems and robots meet on a yearly basis.

1.4 Analysis of Composition of STCRs

1.4.1 Basics of Robot Composition

An overview and introduction to robot composition is given in **Volume 1**, where an overview of definitions, facts, and figures and the evolution of automation and robotics, as well as the relevant state of the art of knowledge (robot composition strategies, robot kinematics, actuators, sensor and process measuring technology, end-effector technology) is provided, and concepts such as modularity, human–robot cooperative manipulation, open source in robotics, and robotic self-organization are introduced. Within each thematic field, automation and robotics are addressed in general as well as construction-specific issues and applications. Furthermore, concepts, robotic technologies, and developments relevant for highly flexible production settings (e.g., inbuilt flexibility and modular flexibility of kinematic main structures and end-effectors, open source, fast reprogrammability, cellular approaches) that allow for product individualization in general and/or industrialized customization in the construction industry (e.g., in off- or on-site factories) are highlighted.

1.4.2 Robot Composition and STCRs

Robot compositions in different fields (e.g., general manufacturing industries, aircraft industry, shipbuilding industry, building component manufacturing, building prefabrication, the STCR field, automated/robotic on-site factories) possess similarities and also significant differences. Similarities can be observed, for example, concerning the increasing utilization of modular approaches and human–machine cooperative approaches. Differences result mainly from the scale and type of materials and components to be handled, resulting in different types of end-effectors and kinematic structures.

The following fields are relevant for setting the robot composition in STCR systems:

1. Working direction of the system (along the facade, from overhead, etc.)
2. Mobility approach (omnidirectional mobile platform, rail guided, fixed, etc.)
3. Kinematic structure (number of degrees of freedom [DoFs], geometric organization of links, etc.)
4. End-effector design (level of inbuilt dexterity, etc.)
5. Modularity of the system (exchangeability of end-effectors, drive units, and other parts)
6. Sensor systems (parameters sensed, local sensors, global sensors, complexity/accuracy/cost of sensors used, etc.)
7. Control mode (remote controlled/supervised, automatic, human–robot cooperative)

Kinematic structure analysis related to STCR studies the motion of different parts of robots to eventually get the desired on-site task done by the end-effector. Different types and the numbers of joints are studied along with links, which are rigid connections between joints. There are two basic kinds of joints in robots: translational (prismatic) and rotational (revolute). STCRs work in a more dynamic environment, in contrast to many other types of robots (such as industrial robots in the general manufacturing industry, which often are fixed and perform repetitive tasks in an SE). Most STCRs are required to move along defined trajectories or with defined areas and thus to utilize some sort of mobility providing mechanism. Examples of kinematic structures (Figures 1.2 to 1.18) are given in the following section for some STCR systems to provide a better understanding of the motion and degrees of freedom for STCRs of different categories.

1.4.3 Symbols and Representations of Kinematic Structure of STCRs

The kinematic diagrams of the robots are drawn partially by following the standard symbols and representations and partially using necessary additional symbols that are self-explanatory.

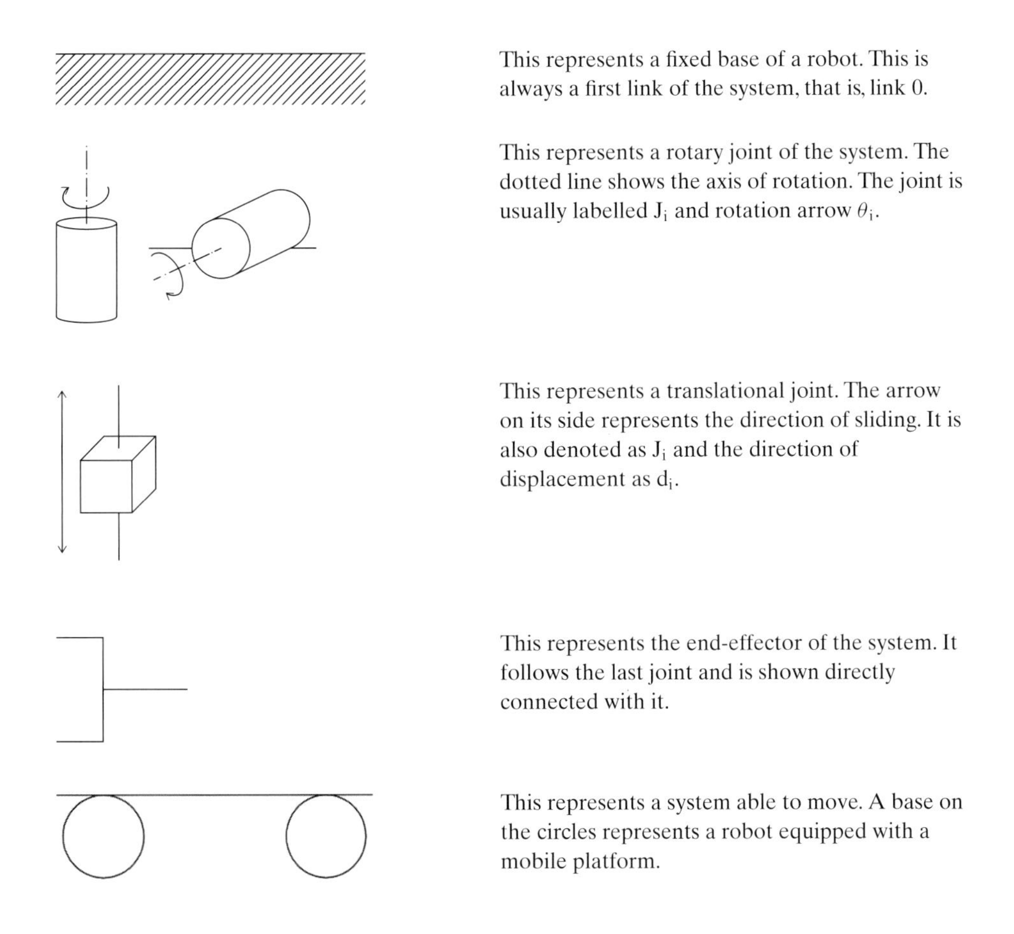

1.4.4 Comparison of Kinematic Structures of STCRs

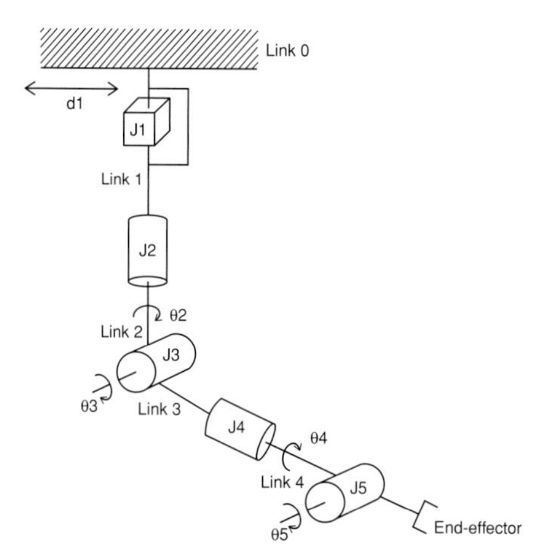

Figure 1.2 Overhead rail-guided digging robot, Shiraishi.

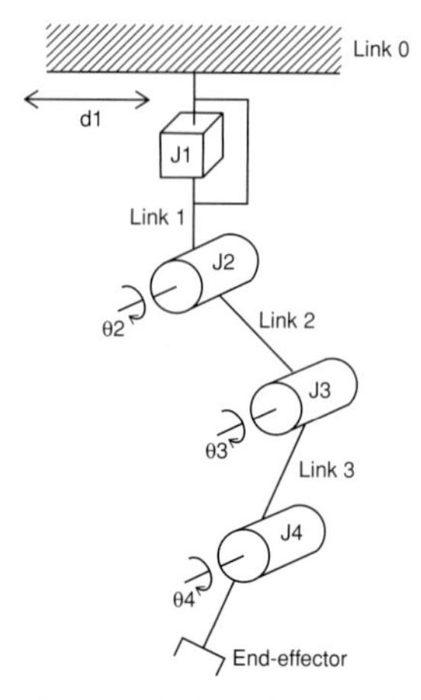

Figure 1.3 Overhead rail guided digging robot, Mitsui.

Earth and Foundation Work

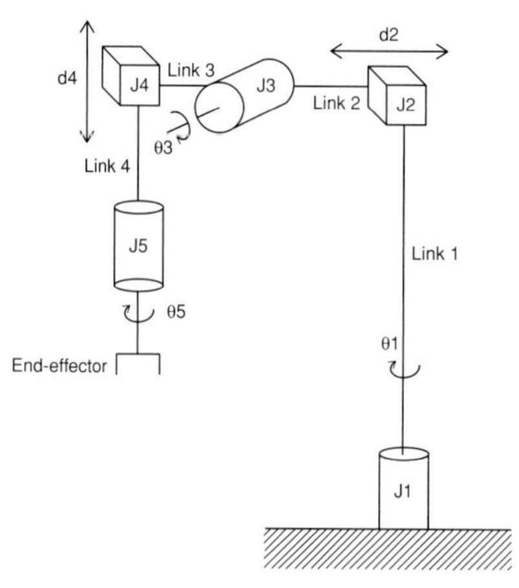

Figure 1.4 Automated crane for rebar positioning, Takenaka.

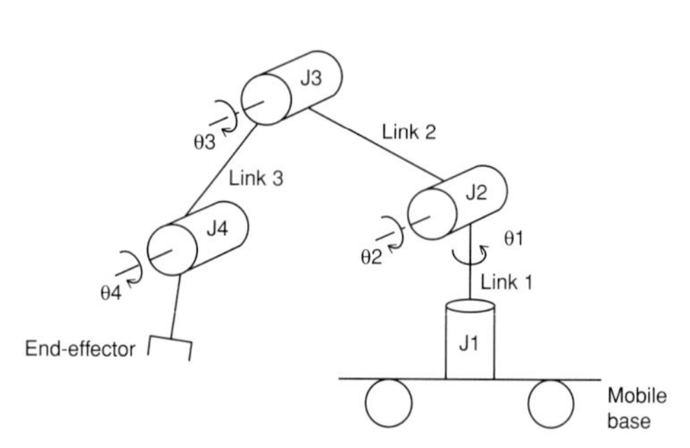

Figure 1.5 Robot for positioning heavy rebar, Kajima.

Reinforcement Positioning

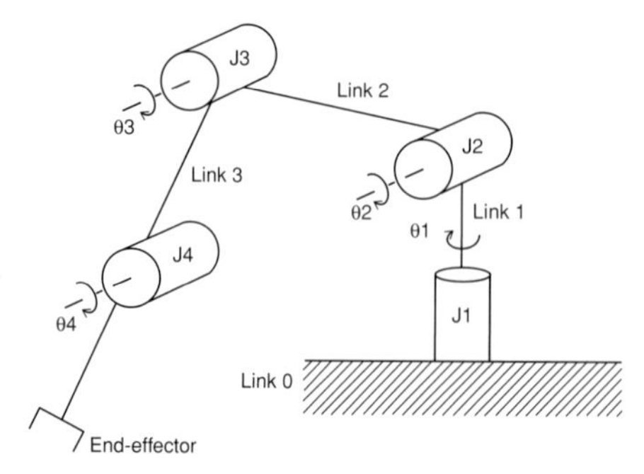

Figure 1.6 Stationary concrete distribution robot, Obayashi and Mitsubishi.

Concrete Distribution

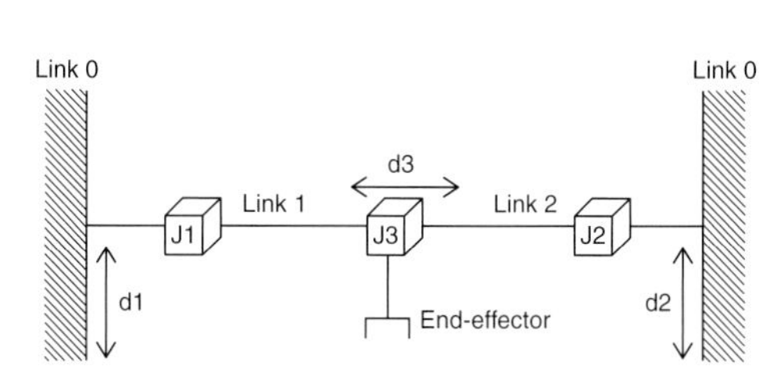

Figure 1.7 Automatic floor screeding robot, Takenaka.

Concrete Levelling

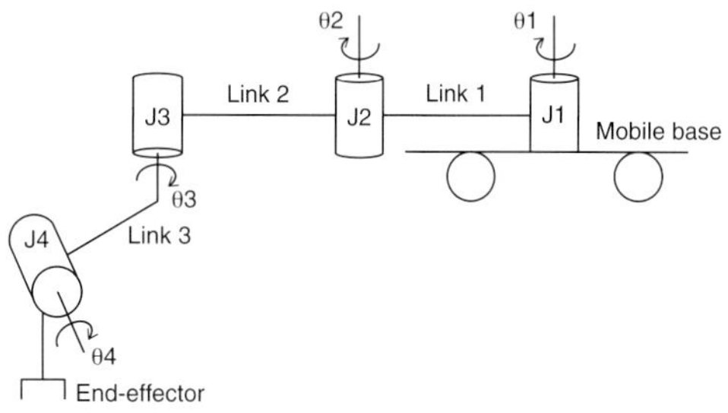

Figure 1.8 Mobile floor finishing robot, Kajima.

Concrete Finishing

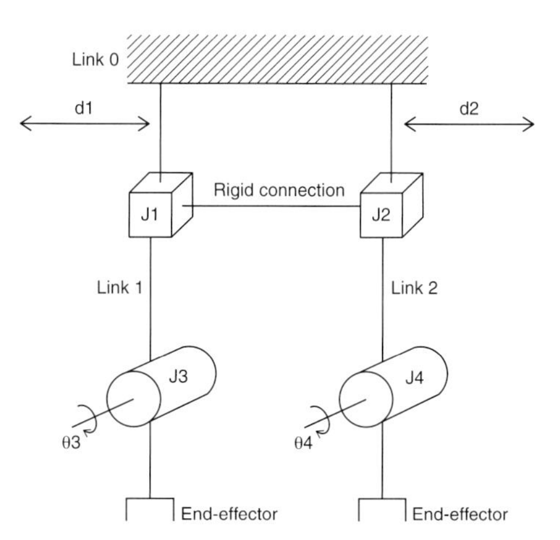

Figure 1.9 On-site monorail overhead logistics, Kajima.

Site Logistics

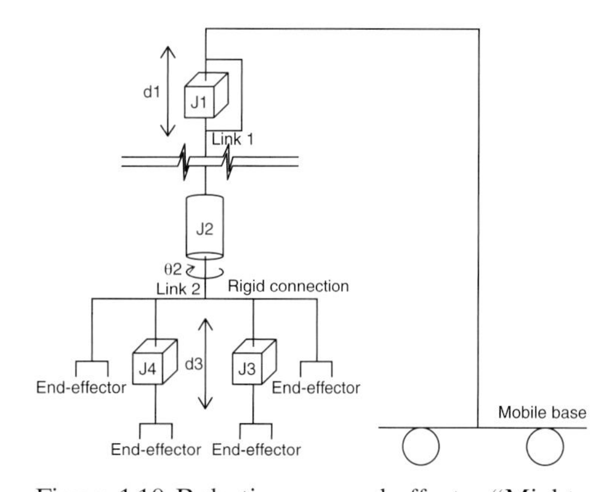

Figure 1.10 Robotic crane end-effector "Mighty Jack", Shimizu.

Crane End-effectors

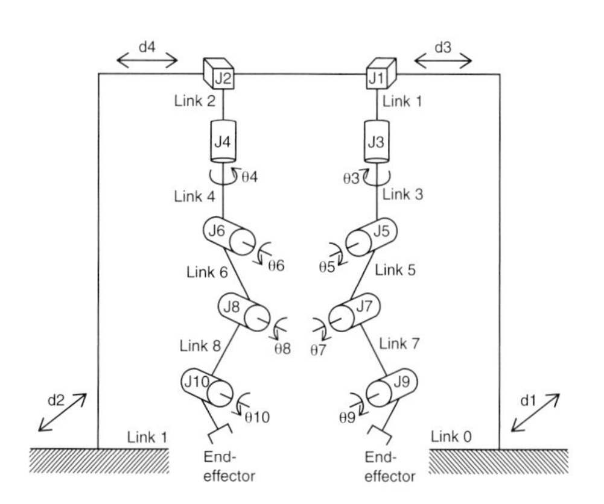

Figure 1.11 Welding machine for steel girders, Kawasaki.

Steel Welding

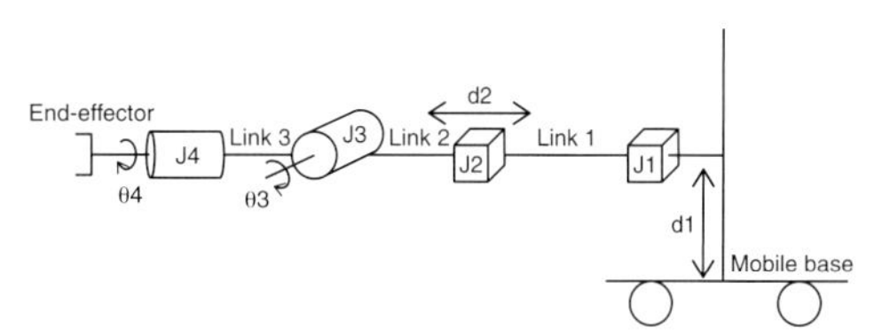

Figure 1.12 Facade element installation robot, Kajima.

Facade Installation

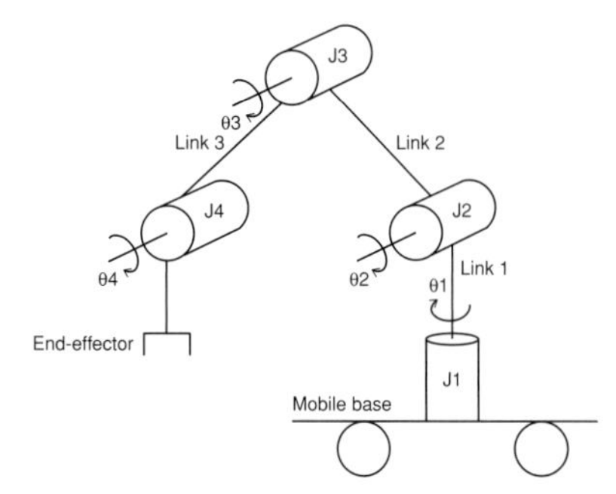

Figure 1.13 Robot for setting tile floor, Eindhoven University.

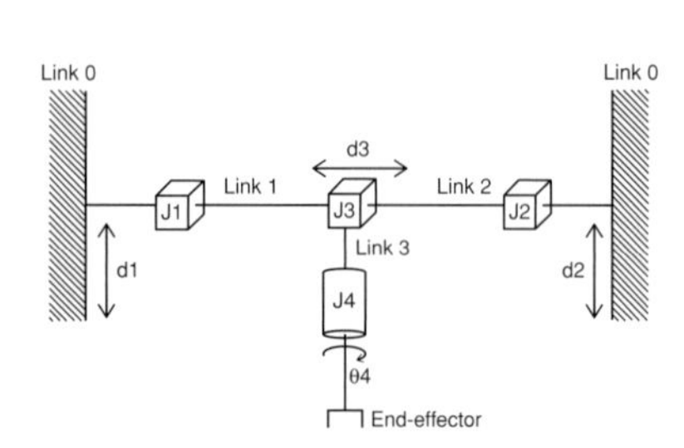

Figure 1.14 Robot for painting exterior walls, Taisei.

Tile Setting

Facade Coating and Painting

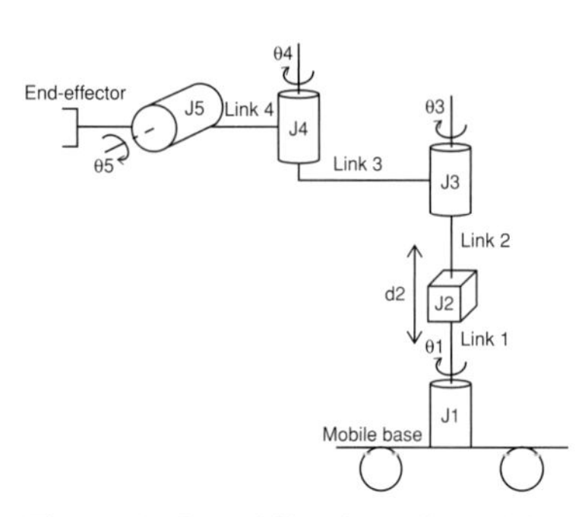

Figure 1.15 Mobile plasterboard finishing robot, Komatsu.

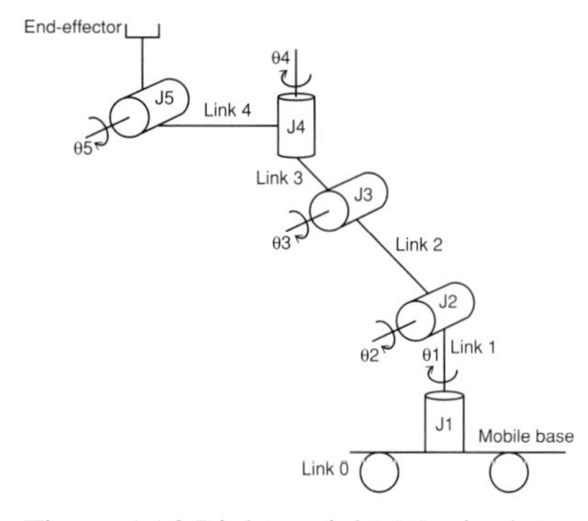

Figure 1.16 Light weight Manipulator for ceiling panel installation, Tokyo Construction Co., Ltd.

Interior Finishing Robots

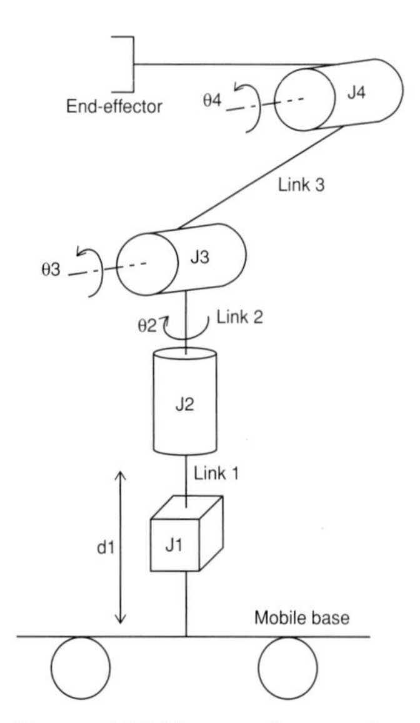

Figure 1.17 Fireproof coating robot, Shimizu.

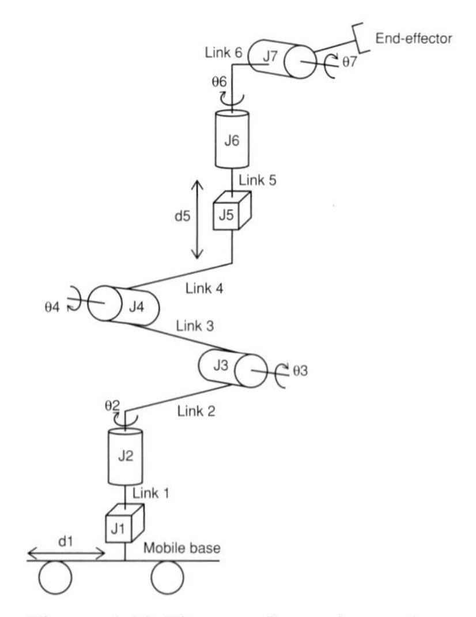

Figure 1.18 Fireproof coating robot, Toda.

Fireproof Coating

STCRs vary in terms of their kinematic structure slightly within each function-oriented category (concrete distribution, painting, fireproof coating, etc.) and also within each category. The most significant differences are in the working direction (the orientation of the robot) and through their moving system and direction. However, the comparison also shows significant similarities of the base bodies of STCRs. For example, STCRs can move on rails, providing them with one DoF as basic motion along a strictly defined route to execute the task (e.g., Taisei robot for painting exterior wall, robotic end-effector by Obayashi, and rail-guided automatic on-ground logistics by Kajima). A variety of construction lifts show a similar mobility approach (e.g., vertical lift by Takenaka and robotic vertical lift by Maeda & Komatsu) as well as some service robot systems (e.g. , Nihon Bisoh's faced cleaning systems).

Systems without rails can be found, for example, in Kajima's mobile floor finishing robot, a facade element installation robot developed by Hanyang University and Samsung, and a robot for positioning of heavy rebar developed by Kajima. In addition, a mobile plasterboard handling robot developed by Komatsu has omnidirectional wheels. Another mobility approach category is represented, for example, by robots suspended by a mechanism or cables from rooftops or ceilings. Some robots are even designed as stationary systems that can be moved only with the help of other system such as cranes (e.g., stationary concrete distribution robot of Obayashi).

1.4.5 Future Research Tasks Related to STCR Robot Composition

The preceding overview of kinematic structures shows that in particular the base bodies of the robots are similar whereas the end-effectors and also the moving system and working directions vary. Usually the base bodies contain a significant amount of actuators and thus account for a large part of the cost of a robot. An intelligent modularization would hold the strongest potential to advance the STCR field. It would allow (similarly to the general manufacturing industry) individual companies to specialize in individual subsystems (mobile platforms, base bodies, end-effectors) and provide a larger variety of solutions for each subsystem in a cost-effective way from which the manufacturer (in the case of construction the construction company, contractors, system developers, or craftsmen) could then choose. This would, however, necessitate development of certain standards and the willingness of individual companies to outsource some parts of the robot subsystem development. Modularity would, furthermore, allow that the same mobile platform, for example, is equipped on the site with different base bodies of end-effectors, which would allow one system on the site to handle over time different tasks during the construction process. This would address the currently very task-specific development and application of the robot system mentioned earlier, which constitutes one of the actual weaknesses of the STCR approach.

2 Single-Task Construction Robots by Category

2.1 Automated Site-Measuring and Construction Progress Monitoring

Robotic systems increasingly are being developed for and used in the context of construction for tasks such as the monitoring of sites, operation and performance of equipment, and the progress of construction (including construction site safety); the surveying and reconstruction of areas, buildings, and facades; and the inspection and maintenance of buildings. Although many robotic systems in this field are still expensive and are just about to transition from the research sphere into the product grade, professional area, the advantages they promise in terms of productivity, coverage of the surveyed area, data richness, speed, work flow and data integration, and last but not least the reduction in human labour needed are enormous. For example, in the context of Irma3D it was shown that the mobile robot can reduce the time surveyors have to spend by 75% (Nüchter et al. 2013). Aerial robots, rapid ongoing improvements in the performance of sensors, and the automation of the whole data acquisition "pipeline" (from motion/flight planning, to registration of multisensor data, the extraction of editable 3D data to the incorporation of the extracted 3D data into building information modeling [BIM] datasets, etc.) are expected to contribute to further significant performance gains. In the future, the implementation of such robot systems will play a major role in the improvement of the overall productivity of construction. In the following sections, an overview of the state of the art of technology in this field is given based on a subcategorization into mobile, ground–based, and aerial approaches. In view of the extensive developments both in the academic and the professional fields in this area, for each category a few examples were selected to be covered in this book to highlight major development directions.

2.1.1 Mobile Robots (Systems 1–4)

Mobile robotic platforms for construction sites can be equipped with a variety of sensors such as laser scanners, thermal inspection systems, and cameras. Increasingly attempts are being made to fuse the information obtained from multiple sensors into one multimodal, coherent image of the construction site, the building, or the area to be inspected or surveyed. In particular, with laser scanners mounted on mobile

	Mobile Robots	
1	Mobile, autonomous robot for real-time gathering of as-built information of construction sites	RICAL – Robotics and Intelligent Construction Automation Laboratory
2	Irma3D – Intelligent robot for mapping applications in 3D	Measurement in Motion, University of Würzburg
3	Robot for the surveillance and surveying of indoor environments	Missouri University of Science and Technology
4	Mobile robotic platforms for measurement and application development	Clearpath Robotics
	Aerial Robots	
5	Octocopter AscTec Falcon 8 equipped with an inspection sensor for synchronized thermal and red, green, blue (RGB) images	Ascending Technologies
6	AL3–32 S1000 Copter equipped with the lightweight light detection and ranging (LiDAR) sensor AL3–32	Phoenix Aerial Systems, Inc.
7	*RIEGL*'s RiCOPTER integrated with a *RIEGL* VUX-1UAV survey grade LiDAR sensor	*RIEGL*
8	Building outline extraction by unmanned aerial vehicles (UAVs) equipped with photogrammetry and LiDAR system	Yang and Chen
9	Inspection of building facades by UAVs equipped with a Kinect sensor	University of Vigo
10	3D facade reconstruction using drones equipped with 3D vision systems	Aerial Vision Group, Graz University of Technology

platforms, highly accurate point clouds and geometric models can be achieved (with *RIEGL* laser scanners, for example, an accuracy of about 5 mm or even better can be attained). However, fusing the information of various sensors into complete 3D models in a highly accurate manner is the subject of ongoing research (see, e.g., Borrmann et al. 2012).

The Robotics & Intelligent Construction Automation Lab (Dr. Yong K. Cho; RICAL 2015) at the Georgia Institute of Technology developed a mobile robot (Figure 2.1) for gathering real-time as-built information on construction sites in the form of 3D point cloud maps with thermal and red, green, blue (RGB) data (Figure 2.2) by means of a laser scanner and thermal inspection system. The mobile platform is also equipped with additional sensors (GPS receiver, lasers, encoders in the wheel) for simultaneous localization and mapping (SLAM) – based autonomous navigation.

The robot Irma3D (Intelligent Robot for Mapping Applications in 3D; Figure 2.3) was developed to conduct build surveying tasks in a highly autonomous manner. Researchers involved in the development showed that the robot can reduce the time surveyors have to spend by 75% (Nüchter et al. 2013). To conduct surveying, the robot is equipped with a 3D laser scanner (VZ-400 RIEGL), a digital single-lens reflex (DSLR) camera (Canon 1000D), and a thermal camera (Optris PI160). Thus the system can produce a point cloud with colour and thermal information. The mobile platform itself is also equipped with several sensors (2D laser scanner, inertial measurement units [IMUs], digital cameras) and allows for both manual

Figure 2.1. Mobile, autonomous robot equipped with several sensors (e.g., 3D laser scanners, thermal cameras) for real-time gathering of as-built information of construction sites. (Image: RICAL – Robotics and Intelligent Construction Automation Laboratory.)

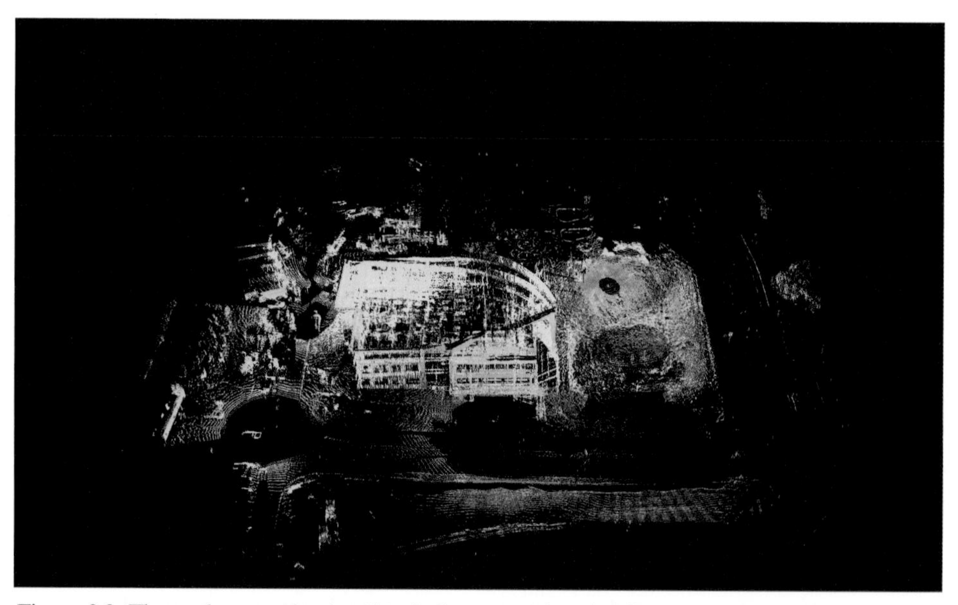

Figure 2.2. Thermal mapped point cloud of construction site. (Image: RICAL – Robotics and Intelligent Construction Automation Laboratory.)

Figure 2.3. Irma3D (intelligent robot for mapping applications in 3D) surveying an historic area for preservation/renovation purposes. (Image: Measurement in Motion, University of Würzburg). The system is equipped with, among other devices, a 3D laser scanner (VZ-400 RIEGL), a DSLR camera (Canon 1000D), and a thermal camera (Optris PI160.)

remote-controlled and autonomous navigation. The robot system was developed by researchers of the Robotics and Telematics group of the University of Würzburg and the Jacobs University Bremen gGmbH as part of a research project. The system is used today by the company Measurement in Motion (a spin-off of the University of Würzburg), which provides professional data acquisitions services.

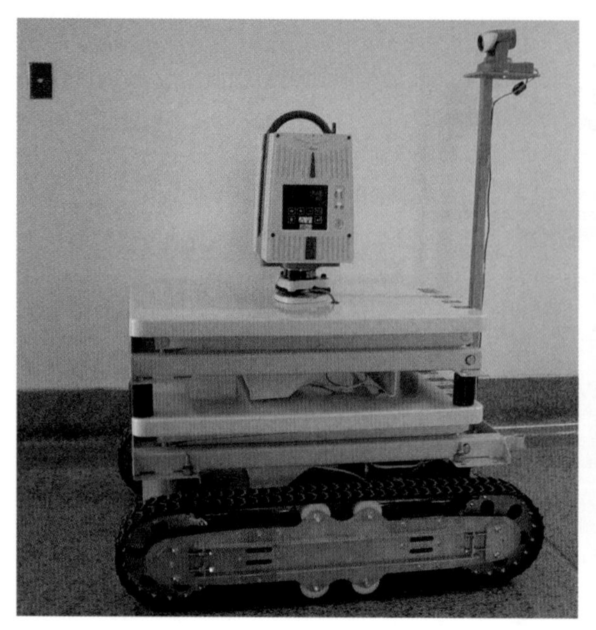

Figure 2.4. Mobile tracked and remote-controlled platform for indoor surveying equipped with several scissor mechanisms to lift and position a sensor system (infrared and LiDAR). (Images: Missouri University of Science and Technology, Prof. Maerz.)

Figure 2.5. Several scissor mechanisms allow lifting and positioning of a sensor system flexibly. (Images: Missouri University of Science and Technology, Prof. Maerz.)

An experimental system for surveillance and surveying (including the detection of cracks and damages, etc.) of indoor environments was developed by the Missouri University of Science and Technology (Department of Geological Engineering, principal investigators: Prof. Dr. Maerz, Prof. Dr. Ye Duan; the project was funded by the Leonard Wood Institute; Figures 2.4 and 2.5). A mobile tracked and remote-controlled platform was equipped with several scissor mechanisms to lift and position a sensor system flexibly. The sensor mechanism consists of a light detection and ranging (LiDAR) system.

The company Clearpath Robotics provides a range of mobile platforms (manually controlled and fully autonomous) professionally integrated with sensors such as stereo cameras, radar, LiDAR, GPS, and IMUs for measurement and application development (Figures 2.6 and 2.7).

2.1.2 Aerial Robots (Systems 5–10)

Unmanned aerial vehicles (UAVs) are increasingly being developed for and used in the context of construction for tasks such as the monitoring of construction processes; the surveying and reconstruction of areas, buildings, and facades; and the inspection and maintenance of buildings. In general, they can be distinguished between low-cost, mini UAVs that can handle payloads of 1–4 kilograms and high-performance UAVs made for use in the professional context with payloads of up to 10 kilograms. Sensor payloads for UAVs have to be extremely light and have to make a tradeoff between their weight and the accuracy they can achieve. Typical sensor payloads for UAVs in

Figure 2.6. Sensing and localization technology professionally integrated by Clearpath Robotics with a mobile vehicle for surveying purposes. (Image: Joshua Marshall, Queen's University.)

Figure 2.7. Advanced Mining Technology Center (AMTC) and the University of Chile working with a 3D laser scanning system based on the HUSKY platform of Clearpath Robotics for mapping a mine autonomously. (Image: Clearpath Robotics, Inc.)

the construction context are sensors for 3D imagery and photogrammetry, thermal cameras/sensors, magnetometers, and LiDAR/laser scanners. The accuracy of UAV-based data acquisition (in particular, 3D data) can be considered situated between that of ground-based approaches (Robotic Total Station, Tachymetry, mobile robots, etc.) and air- and space-borne systems. However, UAVs promise to reduce both cost and effort (in terms of man-hours) significantly, and at the same time they can cover large areas and buildings with minimal effort.

Technically challenging in the context of data acquisition by UAVs is the development of flight planning software that automates (depending on the sensing approach used) the setting up of the flight plan (e.g., as shown by Teizer and Siebert 2014) as well as the equipping of the UAVs with the right combination of flight control units and additional sensors (IMUs, GPS receivers, etc.) that allow an accurate determination of the location of the UAV and the orientation of its data acquisition equipment. Increasingly, approaches are being developed for the fusion and registration of information obtained by differed sensors attached to one UAV. Yang and Chen (2015), for example, have been working on methods of automating co-registration of photogrammetric data and data obtained by LiDAR systems (Figure 2.12). UAVs equipped with photogrammetric systems are able to obtain highly accurate data on existing buildings and building facades (Aerial Vision Group, Graz University of Technology: Daftry et al. 2015; Figure 2.14) that can be used as the basis for the planning and construction/renovation phase. Experiments with low-cost sensors such as the Kinect sensor attached to UAVs are also increasingly being conducted (Figure 2.13).

The German company Ascending Technologies, for example, develops and manufactures high-performance UAVs (Octocopters, Figure 2.8) that can be equipped

Figure 2.8. Octocopter AscTec Falcon 8 equipped with a payload inspection TZ71 sensor for synchronized 14-Bit RAW thermal and RGB images. (Image: © Ascending Technologies.)

and integrated with a variety of sensors such as digital cameras for photogrammetric data acquisition, thermal cameras, and video cameras. Phoenix Aerial Systems (Figure 2.9) and *RIEGL* (Figures 2.10 and 2.11) have both developed their own UAVs, which are able to carry their UAV-oriented, lightweight LiDAR sensors.

Figure 2.9. AL3–32 S1000 Copter equipped with the lightweight LiDAR sensor AL3–32. (Image: Phoenix Aerial Systems, Inc.)

Figure 2.10. *RIEGL*'s RiCOPTER integrated with a *RIEGL* VUX-1UAV survey grade LiDAR sensor. (Image: © *RIEGL*, www.riegl.com)

Phoenix Aerials AL3–32 S1000 Copter has spreadable wings and can be folded into a compact entity for transport in a car. Equipped with the lightweight LiDAR sensor AL3–32, it weighs less than 11 kilograms.

RIEGLs' RiCOPTER maximum take-off mass (MTOM) is specified below 25 kilograms, including a payload of 16 kilograms consisting of the fully integrated VUX-1UAV survey grade LiDAR sensor, an IMU, additional sensors (e.g., different types of cameras), and the necessary batteries. With the *RIEGL* VUX-1UAV survey grade LiDAR, sensor data acquisition is accurate to about 10 millimetres.

Figure 2.11. Perspective point cloud view of an area obtained with a RiCOPTER. (Image: © *RIEGL*, www.riegl.com)

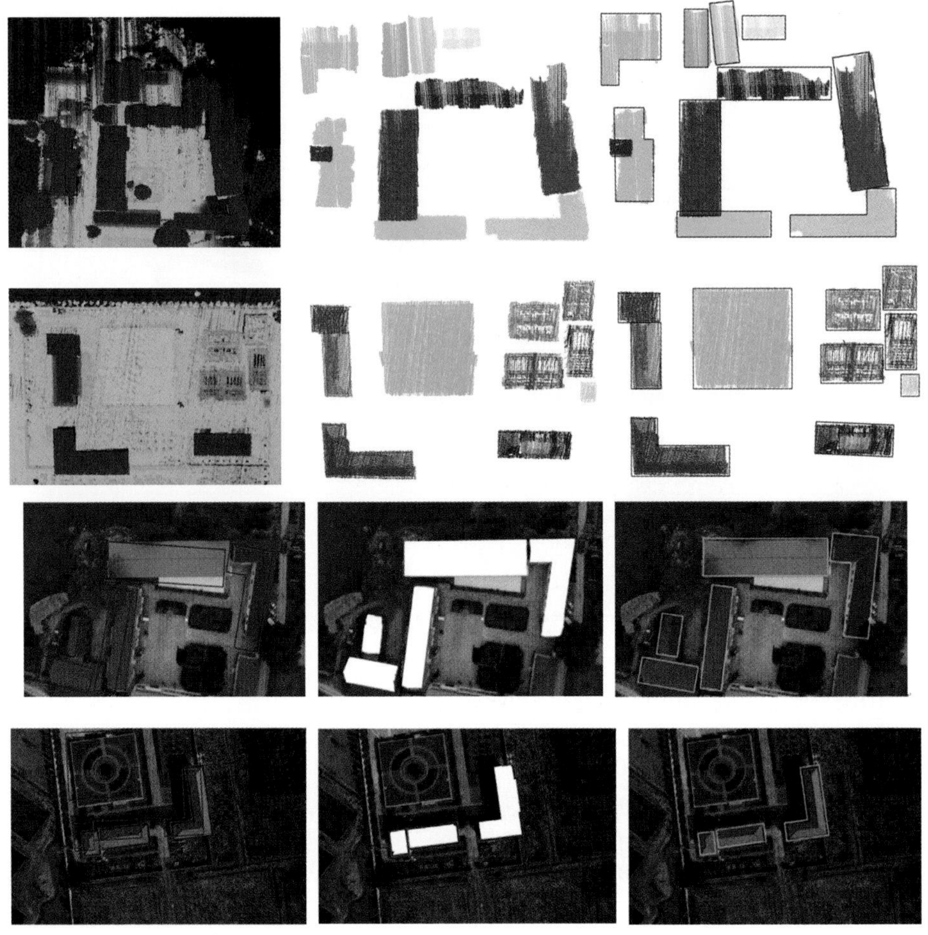

Figure 2.12-1. Highly accurate extraction of building shapes by photogrammetry based on prior knowledge acquired from LiDAR data. (Reprinted from Yang and Chen 2015, "Automatic registration of UAV-borne sequent images and LiDAR data," *Journal of Photogrammetry and Remote Sensing*, 101:262–274, with permission from Elsevier.)

Figure 2.12-2. UAV equipped with camera/photogrammetry and LiDAR system. (Yang and Chen 2015; image: Chen.)

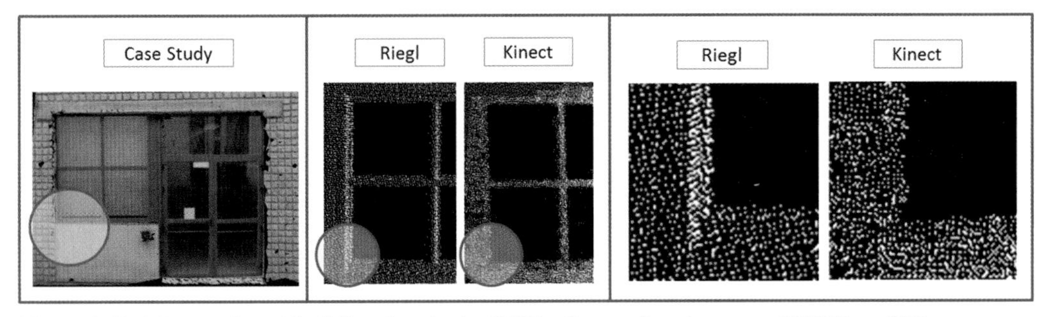

Figure 2.13-1. Inspection of building facades by UAVs. Comparison between *RIEGL* and Kinect point clouds. (Reprinted from Roca et al. 2013, "Low-cost aerial unit for outdoor inspection of building facades," *Automation in Construction*, 36:128–135, with permission from Elsevier.)

Future research in the field will focus more and more on the automation of the whole data acquisition "pipeline" from flight planning to registration of multisensor data, the extraction of editable 3D data (planes, 3D volumes, etc.) from the raw sensor data (e.g., point clouds) to the incorporation of the extracted 3D data into BIM models (also referred to as "scan-to-BIM"; see, e.g., Xiong et al. 2013; Wang et al. 2015).

2.2 Earth and Foundation Work Robots

Earth and foundation work is an essential component of every type of construction process. Furthermore, in contrast to other types of work related to the construction of elevated sections of a building, earth and foundation work has a limited impact on a building's individuality. Thus, it is a work process that, seen over several projects, appears in considerable measure, and at the same time it is a process in which many repetitive subactivities occur (e.g., excavation operations). Above all, the work associated with this category is considered to be among the most dangerous in construction. As this forms an ideal basis for automation, the approaches in that field are numerous, ranging from robots for assisting in the on-site production of

Figure 2.13-2. Inspection of building facades by UAV equipped with a Kinect sensor. (Roca et al. 2013; image: Applied Geotechnologies Research Group – University of Vigo.)

Figure 2.13-3. Inspection of building facades by UAV equipped with a Kinect sensor. (Roca et al. 2013; image: Applied Geotechnologies Research Group – University of Vigo.)

Figure 2.14. 3D facade reconstruction using drones equipped with 3D vision systems. (Daftry et al. 2015; image: Aerial Vision Group, Graz University of Technology.)

individual diaphragm walls to robots for automated excavation (ground-based and overhead solutions) and automated soil removal systems. More complex solutions in this category not only allow performance of one operation but also try to build up interconnected automated chains (e.g., integrated and automated soil loosening, excavation, and soil removal process).

Earth and Foundation Work Robots		
11	Overhead rail-guided digging robot	Various developers; for example: Shiraishi, Daiho Construction, Advanced Construction Technology Center, Ministry of Construction, Ohmoto Gumi, etc.
12	Automatic excavation system for diaphragm-wall production	Konoike
13	Automatic digging and soil removing system	Tokyu Construction Co., Ltd.
14	Overhead rail-guided digging robot	Mitsui
15	Automatic on-site soil removing/logistics system	Kajima
16	Em 320 S robot for digging vertical shafts	Shimizu

System 11: Overhead Rail-Guided Digging Robot

Developer: Various developers; for example Shiraishi, Daiho Construction, Advanced Construction Technology Center, Ministry of Construction, Ohmoto Gumi, and so forth.

Work tasks performed at great depths encounter various problems such as high atmospheric pressure in the working environment. Use of unmanned and semi-automated overhead rail-guided digging robots addresses this problem (Figure 2.15). All necessary information about the system as well as the work process is displayed in the operation control room in real time, and the operators can oversee the excavation process while remaining above ground. The following are features of the tele-operation strategy:

- The caisson shovel is controlled remotely by an operator.
- The loading operation by the shovel, which is a simple repetitive process, is automatically controlled.
- The soil loader operations are automatically controlled.
- Operation of the whole system can be monitored in real time because sensors are incorporated into all subsystems.

In some cases the automated excavation system is combined with systems for automated soil removal (automated soil logistics from the excavation area to the ground/disposal area) and systems for systemized automated production of the basement.

For further information, see Council for Construction Robot Research (1999, pp. 280–281) and Kodaki et al. (1996).

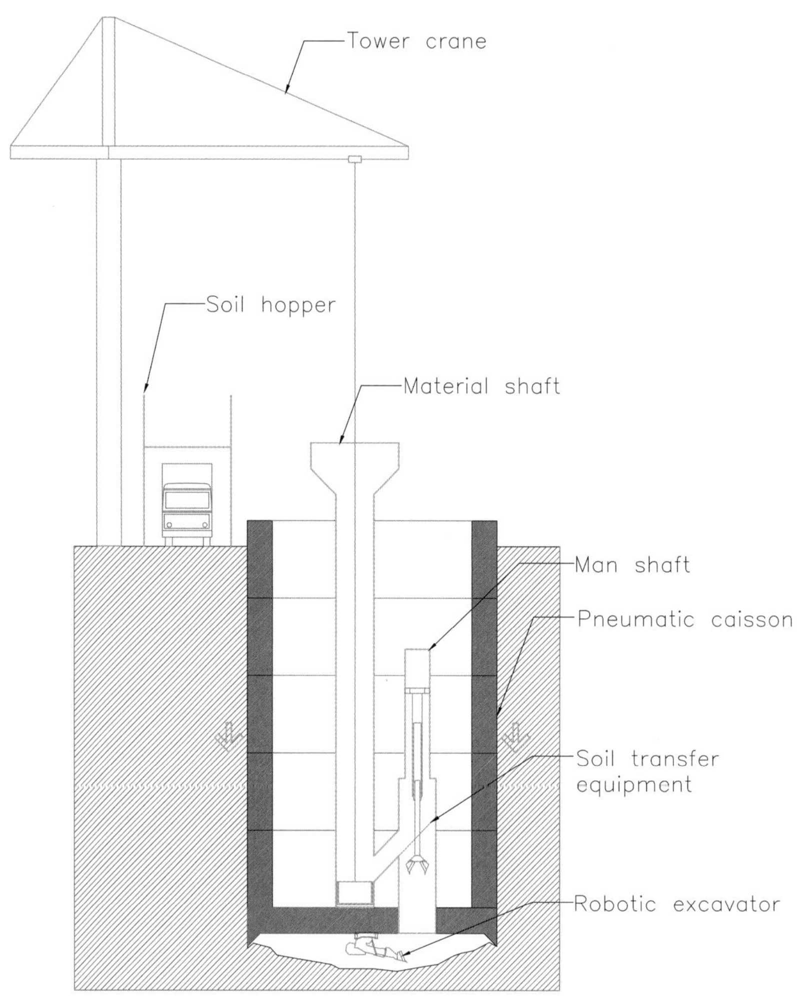

Figure 2.15. Overview of the total overhead rail-guided digging robot system. (Graphical representation according to Kodaki et al. 1996.)

System 12: Automatic Excavation System for Diaphragm-Wall Production

Developer: Konoike Construction Co. Ltd. and Faculty of Engineering, Ehime University

The system (Figure 2.16) is composed of various subsystems: an excavator load control system and an excavator positioning system. The excavator is automatically positioned by an integrated control system and is equipped with measuring and guidance equipment consisting of an inclinometer, a depth meter, a personal computer, and a fuzzy controller. The excavation load control system controls the downward speed of the excavator as well as the exertion of force on the soil. The system is capable of achieving a tolerance of 30 millimetres along the vertical excavation line, and in case of deep excavation, it maintains an accuracy of 30–50 millimetres at a depth of 100 metres. All of the data collected and processed by different sensors and systems installed at the excavator, base machine, and measuring devices are available for display on the monitor in the control cabin.

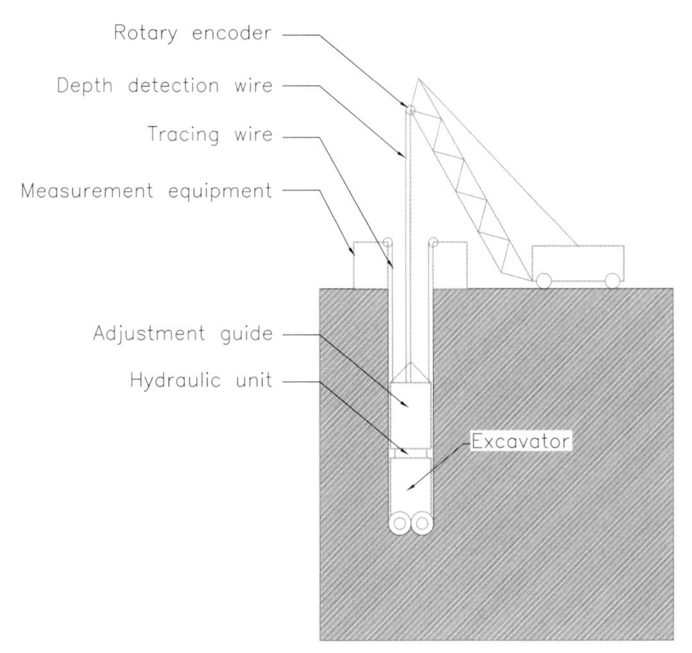

Figure 2.16. Automatic excavation system for diaphragm-wall production. (Adopted from Matsuike et al., 1996.)

For further information, see Council for Construction Robot Research (1999, pp. 16–17) and Matsuike et al. (1996).

System 13: Automatic Digging and Soil Removing System

Developer: Tokyu Construction Co., Ltd.

This robot was developed to improve the working environment for deep excavation operations and to relieve human workers from this hazardous task. Excavation workers face risks of injury from collapsing sides of a deep pit, improper air ventilation, and exposure to toxic gases that might be produced by construction machinery. They may also come across the problem of groundwater infiltration from the bottom or falling objects from the top of the pit. In the automatic digging and soil removing system, a skilled worker can operate the digging equipment from a safe distance by controlling it remotely. An emergency shutdown switch is provided to stop all functions of the robot in case of any emergency or safety issue. The system uses a compact digging robot with a size of 1.75 m × 1.19 m, and weighing only 2130 kilograms as the central element (Figure 2.17). It is particularly suitable in situations in which little working space is available. In addition, the system utilizes an automated lifting system: the digging robot places the soil in a loading compartment (Figure 2.18). The digging robot contains two manipulators. One manipulator is equipped with an end-effector for loosening the soil material from the ground (cutter drum mechanism; Figure 2.19). The second manipulator is equipped with an excavation bucket (with a capacity of 0.025 $m^{3)}$ to remove the loosened soil material and to fill the loading compartment.

For further information, see Council for Construction Robot Research (1999, pp. 30–31).

Figure 2.17. Excavation robot in action. (Image: Tokyu Construction Co., Ltd.)

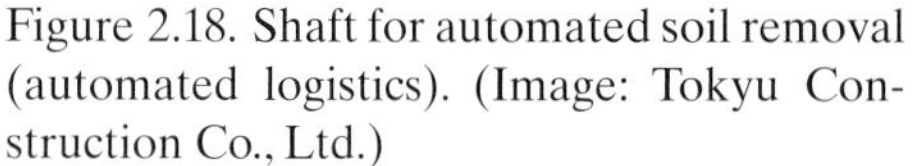

Figure 2.18. Shaft for automated soil removal (automated logistics). (Image: Tokyu Construction Co., Ltd.)

Figure 2.19. End-effector for soil loosening. (Image: Tokyu Construction Co., Ltd.)

System 14: Overhead Rail-guided Digging Robot

Developer: Mitsui

A key element of this experimental system is a hydraulically activated overhead manipulator with a kinematic structure specially designed for conducting digging or excavation operations (Figure 2.20). The manipulator can be equipped with a variety of end-effectors (shovels) to perform the digging/excavation process. The manipulator can be installed to overhead rail systems or movable gantry cranes that can then move it across the work area. The system was conceived as a multipurpose system that can be applied to a large variety of operations, which allow the installation of an overhead frame or system that allows the robot to be installed and moved on the site.

System 15: Automatic On-site Soil Removing/Logistics System

Developer: Kajima

When it is necessary to deal with large-scale projects, the utilization of conventional methods is not efficient. The automatic on-site soil removing/logistics system was introduced by Kajima for large-scale excavation projects in the context of the construction of underground water tanks. It transports excavated soil from a depth

Figure 2.20. Overhead rail-guided digging robot system in action.

Figure 2.21. Automatic on-site soil removing/logistics system.

of more than 35 metres, and in large volume in a semi-automatic manner. The maximum possible depth this system can operate at is 100 metres. The system is composed of several subsystems: a frame system consisting of a vertical and a horizontal part, a logistics system allowing transport of containers and soil through the frame from the excavation area to the ground level area, and various standardized containers (Figure 2.21). The soil-moving machines (excavators, bulldozers, etc.; can be robotic or conventional) move the soil to containers that are moved into the frame when filled completely. From there the soil is transported through the frame partly through a conveyor belt system to the ground level, where the soil is either deposited or removed by haulers. In parallel, the system allows containers to move through the frame to allow the transport of construction materials to the excavation/work area. Furthermore, the system allows efficient transport of personnel to the work area.

System 16: EM 320S Robot for Digging Vertical Shafts

Developer: Shimizu Corporation

This excavation robot is a subsystem of the Shimizu cast *in Situ* Substructure System-Giant (SSS-G), and is used for deep excavation with unconventional dimensions. The system has applications for large structure works such as dams, foundations of bridges and piers, underground tank walls, and so forth. It is a high-performance digging robot for vertical shafts and can excavate thicknesses of up to 3.2 metres at a depth of 150 metres. In addition to the EM-320S excavation robot, the whole system of the SSS-G for deep construction consists of a special scaffold with built-in automatic excavation control, a heavy-duty earth separation and recycling system, and an automatic slurry component analyser system (Figure 2.22). The EM-320S digging

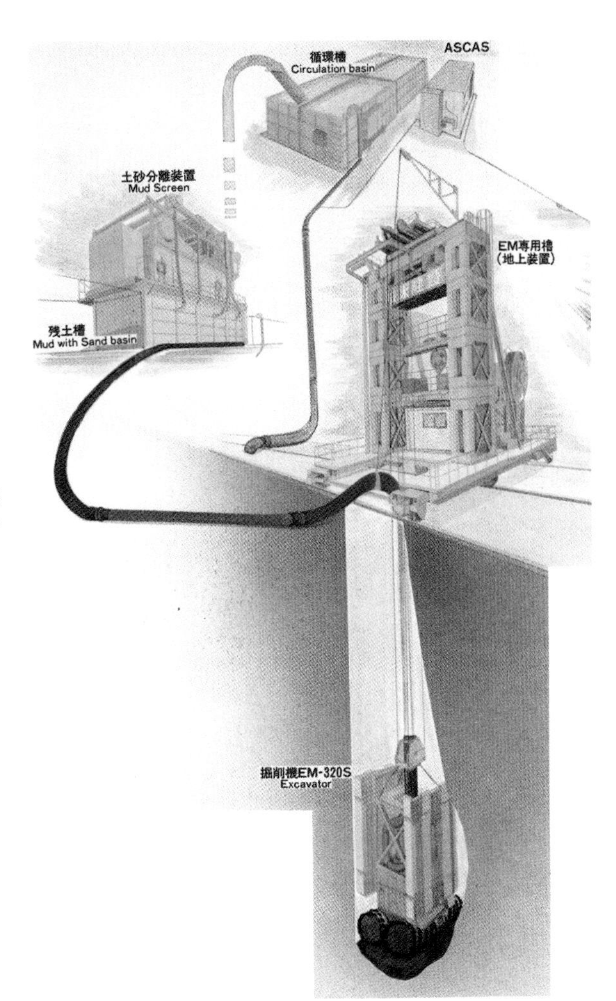

Figure 2.22. EM 320S robot for digging vertical shafts. (Image: Shimizu Corporation.)

robot has a weight of 45 tons and is composed of excavating bits of four drum-cutters, two ring-cutters, and two wing-bits. It can be applied to different soil types such as alluvium, diluvium, and other formation/rocks with high excavation accuracy.

2.3 Robotized Conventional Construction Machines

In contrast to the development of new and highly task-specific construction robots that would also radically transform the way buildings are designed and constructed, the upgrading of currently used "multipurpose" equipment and machines (dumpers, haulers, excavators, etc.) to function as robotic systems is considered a feasible alternative. This approach is particularly attractive because it could provide firms a solution that could be implemented in the short run, building on former models and investments. It would require relatively low development effort, without necessitating major changes in the overall construction process and the present skills of construction workers. Robotized conventional construction machines would provide high task flexibility, intelligently assist human beings, automate a broad range of tasks, and integrate and digitally communicate seamlessly with other machines and tasks.

In the following section, an overview of the state of the art of technology in this field is given. In view of the extensive developments both in the academic and the professional fields in this area, a few examples (concepts and experiments as well as commercially deployed systems) were selected to provide an overview of the major development directions.

Robotized Conventional Construction Machines		
17	Volvo Construction Equipment concept machines	Volvo CE
18	Autonomous construction equipment	Komatsu
19	Concept for a robotic dumper	Wacker Neuson and FH Joanneum
20	Add-on modules for upgrading conventional construction vehicles with robotic functionality	Autonomous Solutions, Inc.
21	AutoMine product family	Sandvik

System 17: Volvo Construction Equipment Concept Machines

Developer: Volvo Construction Equipment (Volvo CE)

Volvo CE has developed a series of innovative concept machines that allowed the company's engineers and designers to challenge the way construction equipment is used. The futuristic articulated hauler concept – the Centaur (Figure 2.23) – will be highly modular, allowing one to quickly switch the functionality of the machine. The hauler will be a quiet, electric machine that will minimalize the disturbance of construction work. The electric motors will be embedded into each wheel, allowing the wheels to move freely over rough surfaces. With the GaiaX (Figure 2.24), Volvo presents a concept for a future compact, lightweight excavator. The concept puts the operator in the centre, optimizing ergonomics and control. As well as operating the machine in the conventional way, the excavator can be controlled remotely with a tablet computer. The Gryphin (Figure 2.25) is Volvo's futuristic concept model for an extreme wheel loader. The electric motors are integrated into each wheel – allowing for independent, three-dimensional movement of the wheel elements, making the machine extremely flexible and suitable for almost any terrain.

Figure 2.23. Centaur: Volvo CE's concept for a futuristic modular, electric articulated hauler. (Image: Volvo Construction Equipment.)

Figure 2.24. GaiaX: The concept for Volvo CE's futuristic lightweight, compact excavator. (Image: Volvo Construction Equipment.)

Figure 2.25. Gryphin: Volvo CE's concept for an extreme wheel loader of the future. (Image: Volvo Construction Equipment.)

Figure 2.26. Intelligent blade control. (Image: Komatsu Ltd.)

System 18: Autonomous Construction Equipment

Developer: Komatsu Ltd.

Komatsu transferred its experience with autonomous single-task construction robots (STCRs) to its conventional construction machinery and managed to successfully commercialize conventional construction equipment upgraded with autonomous functionality. Komatsu has equipped its D61EXi/PXi-23 with an intelligent blade control system as a standard functionality (Figure 2.26). Further equipment incorporating this system will follow. A Global Navigation Satellite System antenna, an enhanced IMU, and stroke-sensing cylinders allow the machine to create an image of the condition on-site and intelligently optimize the movement of the blade to enhance performance and minimize damages and the number of maintenance operations required. Komatsu has with its Autonomous Haulage System (AHS; Figure 2.27) developed one of the world's first commercialized autonomous construction

Figure 2.27. Autonomous Haulage System. (Image: Komatsu Ltd.)

Figure 2.28. The COMB system in operation. (Image: FH JOANNEUM Industrial Design, Graz.)

equipment systems. The system has been tested since 2008 in field trials in cooperation with Rio Tinto. In the context of Rio Tinto's concepts for the mine of the future, it is planned that several hundred autonomous haul trucks will be deployed in Australian mines.

System 19: Concept for a Robotic Dumper

Developer: Wacker Neuson and FH Joanneum

Wacker Neuson, a German manufacturer of construction equipment, cooperates with the University of Applied Sciences FH Joanneum, Department for Industrial Design, for the design and development of future equipment for movement of earth. As part of a design course a student team (R. Akselrud, P. Schütz), under the guidance of Prof. M. Lanz and K. Hilgarth, designed the *COMB* system. The system consists of a robotic mobile platform (Figure 2.28) and separate loading platforms (Figure 2.29). The robotic mobile platform is able to take up, move, and unload the loading platforms (Figure 2.30). The separation into two elements allows the mobile robotic platform to be operated continuously while individual loading platforms rest at certain locations where they are filled with soil or other material.

System 20: Add-On Modules for Upgrading Conventional Construction Vehicles with Robotic Functionality.

Developer: Autonomous Solutions, Inc.

Autonomous Solutions, Inc. (ASI) is a US developer and manufacturer of retrofitting kits that convert conventional construction vehicles to unmanned robotic

Figure 2.29. Mobile, robotic platform taking up a loading platform. (Image: FH JOANNEUM Industrial Design, Graz.)

Figure 2.30. Unloading process. (Image: FH JOANNEUM Industrial Design, Graz.)

vehicles (Figures 2.31–2.33). ASI's automation kit (Figure 2.34) consists of both hardware components such as a robotic steer ring, activators, and a large variety of sensors as well as the command and control software Mobius®. The customer can choose from a variety of modules and functions that allow one to customize the level of automation for each vehicle. The control software, Mobius®, provides an easy-to-use user interface that allows one operator to oversee an entire fleet of vehicles. ASI makes its automation kit available for purchase to its customers and for those without the necessary knowledge also offers integration for their kit into the equipment. ASI's kit can be used to upgrade construction vehicles such as wheel loaders, trucks, dozers, and excavators to unmanned vehicles.

Figure 2.31. Articulated dump truck with ASI's autonomous vehicles kit. (Image: ASI.)

Figure 2.32. Autonomous dozer upgraded by ASI. (Image: ASI.)

Figure 2.33. Excavator upgraded by ASI. (Image: ASI.)

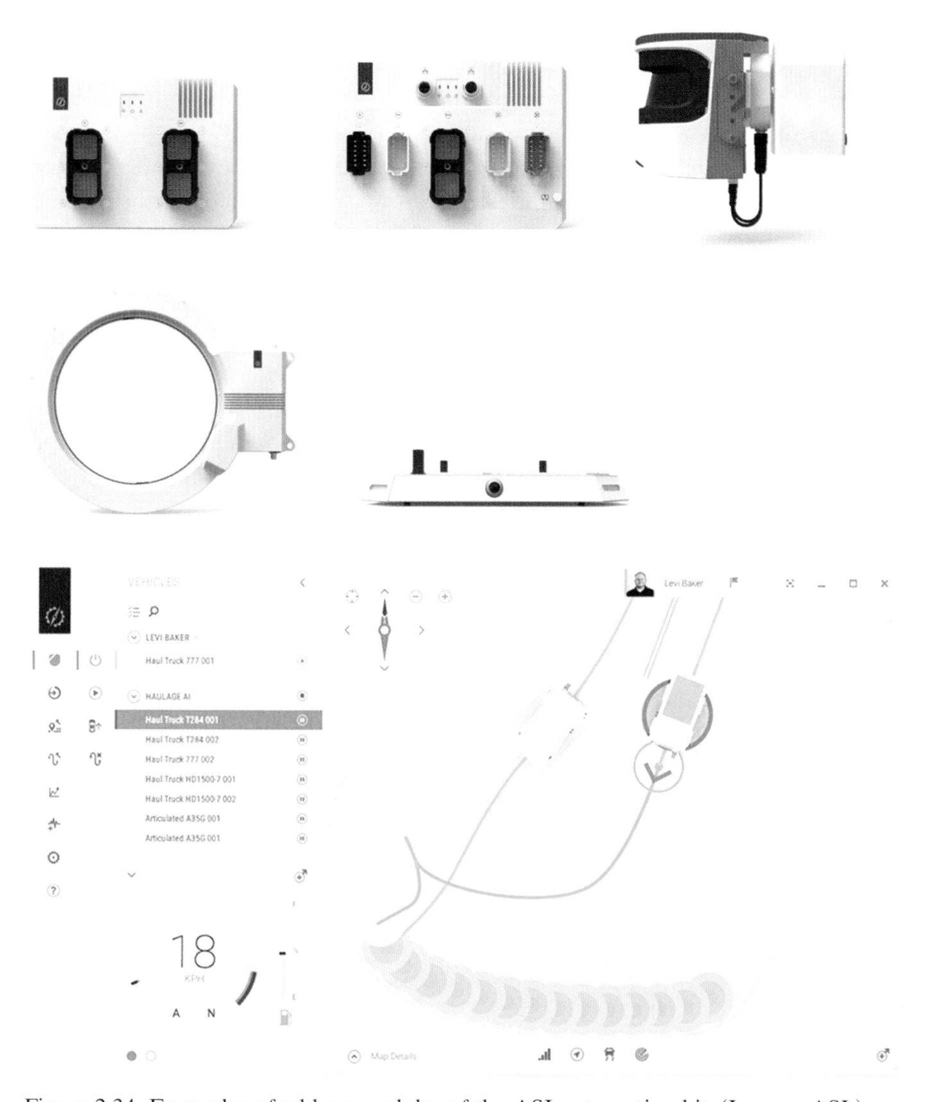

Figure 2.34. Examples of add-on modules of the ASI automation kit. (Images: ASI.)

Figure 2.35. Automated loader LH514. (Image: Sandvik Mining.)

Figure 2.36. OptiMine production control room. (Image: Sandvik Mining.)

System 21: AutoMine Product Family

Developer: Sandvik Mining and Construction

Sandvik's AutoMine product family provides modular elements such as equipment, sensor systems, control room equipment, and software for the automation of earth movement tasks in mining and construction (Figures 2.35–2.39). Depending on the individual use case the system can be adapted to the requirements and allow for the automation of individual tasks in the "lite" version (e.g., automation of a loading operation at a certain location) up to the "fleet" version (e.g., automation of a whole fleet of equipment). Automated equipment is supervised from a control station and it can be operated in fully automated mode as well as in a remote-controlled manner. Laser scanners and camera systems mounted onto the equipment generate the informational basis for automation and operators. Furthermore, beyond individual control stations, Sandvik provides software and equipment for whole production control rooms from where one or several mines or sites can be supervised, and the operation of the equipment can be planned and optimized. Sandvik's equipment monitoring and production management software can be interfaced with other enterprise resource management systems.

Figure 2.37. AutoMine equipment control station. (Image: Sandvik Mining.)

Figure 2.38. AutoMine equipment control station. (Image: Sandvik Mining.)

Figure 2.39. OptiMine task management personal digital assistant. (Image: Sandvik Mining.)

2.4 Reinforcement Production and Positioning Robots

Operations related to reinforcement production and positioning include the cutting and bending reinforcing bars, the precise (relative to each other) arrangement of those reinforcement bars, the binding of reinforcing bars, and the final positioning of the reinforcement bar element or mesh on a floor, in a mold, or in a formwork system. Damage to a human worker's musculoskeletal system is significantly more likely during reinforcement work than, for example, during painting work (Wickström et al. 1985). Furthermore, ergonomically critical postures, such as lifting and carrying operations, represent a severe health threat in the long term. For health insurance companies, the conventional labor-based operations related to reinforcement and reinforcing bar production and positioning are therefore considered as high-risk work both in the short and long term. Automated systems mitigate these risks and the impact on the health of the workers and at the same time enhance the productivity and quality of work related to reinforcement production and positioning. The systems developed for this purpose include those for bending/shaping of individual reinforcing bars on the construction site and the interconnection of reinforcing bars to larger reinforcement elements and meshes on the site. Compared to robots and machines used in the factory to preconfigure reinforcement elements, the systems used on the site need to be highly mobile and compact to suit the purpose of temporary deployment. Furthermore, the category includes smaller sized, mobile robots that can assist workers on the individual floors to handle, position, and fix (medium and heavy weight) reinforcement elements as well as larger, stationary systems for reinforcement delivery and positioning automation.

Reinforcement Production and Positioning Robots		
22	Robot for positioning of heavy reinforcing bars	Kajima
23	Reinforcing bar fabrication robot	Taisei
24	Robot for on-site shaping of reinforcing bars	Obayashi
25	Automated crane for reinforcing bars positioning	Takenaka
26	Robot for positioning of heavy reinforcing bars	Kajima
27	Robot for positioning and installation of medium weight reinforcing bars	Kajima

Figure 2.40. Robot for positioning of heavy reinforcing bars. (Image: Kajima Corporation.)

System 22: Robot for Positioning of Heavy Reinforcing Bars

Developer: Kajima Corporation

For heavy concrete work such as large concrete foundations, heavy reinforcing bars need to be placed and arranged properly at specific locations before pouring concrete. It is difficult for construction workers to carry, manipulate, or fix these heavy reinforcing bars on uneven surfaces. It becomes particularly difficult when more than one steel reinforcing bar mesh is required and must be placed above another for thick concrete placement. For example, a round steel 40-millimetre-diameter bar having a length of 10 metres is difficult to handle manually by workers. The robot consists of a tracked mobile platform and on-board storage for carrying a set of steel bars, a manipulator with 5 degrees of freedom (DOF), and an end-effector for handling the steel bars (Figure 2.40). The length of the end-effector is adjustable to be able to handle a large variety of steel bars of different lengths. The robot can carry about 20 reinforcing bars through the on-board storage and place them one by one at the required locations. It takes approximately 1 minute to place one bar at its designated location.

System 23: Reinforcing Bar Fabrication Robot

Developer: Taisei Corporation

This robot was developed to fabricate steel reinforcing bars on site. It expedites reinforcing bar production and saves on required manpower for the task. It is not,

Figure 2.41. End-effector for reinforcing bar fabrication robot. (Image: Taisei Corporation.)

however, fully automated and there are a few steps that need to be completed manually. Required manual tasks include the cutting and bending of longitudinal bars and stirrups as well as laying down the top longitudinal bars. Following this, the robot (Figures 2.41 and 2.42) takes over the job until the fabrication of a steel reinforcing bar is complete and must be picked up manually. The robot (dimensions: H 1.70 m × L 8.50 m × W 1.50 m) marks the intersection of stirrups on longitudinal bars, places the stirrups at the designated locations and spacings, and secures them with binding wires. The system then locates the bottoms of the longitudinal bars and secures them by binding wire at the locations marked on stirrups. Now, the steel reinforcement is ready to be picked up manually and can be placed in a precast formwork to pour concrete. In contrast to conventional stirrups, which have open-ended hooks, this system uses closed-ended stirrups for automated flash butt welding to maintain bending quality. The robot is able to process a large variety of reinforcement dimensions within its workspace (reinforcement unit dimensions that can be processed: depth: 0.30–0.50 m, length: 4.00–6.00 m, width: 0.25–0.40 m) and needs about 4 seconds per intersection for binding.

For further information, see Council for Construction Robot Research (1999, pp. 264–265).

Figure 2.42. View of total reinforcing bar fabrication robot system. (Image: Taisei Corporation.)

System 24: Robot Systems for On-Site Shaping of Reinforcing Bars

Developer: Obayashi Corporation

Description: The Obayashi robot for on-site shaping of reinforcing bars was intended mainly for the on-site production of reinforcement. The robot was developed to shape larger diameter steel bars by bending them 10°–20° at six possible locations (Figure 2.43). This is called the "Six linked X type reinforcing bar automatic bender unit". Another robot unit is used for column reinforcing bar fabrication at the surface that can produce up to 10-metre-long columns with only two workers. The column reinforcing bar fabrication unit has five linked support arms, a base, hydraulic units, a control unit, and two jib cranes for easy handling of heavy steel bars. Once the input data for reinforcing bar shaping have been entered in stroke setter at the bender unit, all six hydraulic benders of the X type reinforcing bar bending unit work simultaneously. As soon as this bending process is completed, a measurement check is performed for the angles and the length of the reinforcing bar to confirm accuracy. In case of any discrepancy such as a bending angle that is not the same as designed the individual hydraulic bender is used to rectify the error. If no adjustment is required, the reinforcing bar is removed and the benders are reset to their initial position. This reset operation is performed by a simple push button. The same process is repeated for the next reinforcing bar to be shaped. This system was developed for reinforcement work in high-rise buildings. It can, however, bend only within its limit of 10°–20°,

Table 2.1. *Specifications of Robot Systems for On-Site Shaping of Reinforcing Bars*

Six Linked X Type Reinforcing Bar Automatic Bender Units		Reinforcing Bar Column Fabrication Unit at Surface	
Item	Specification	Item	Specification
Hydraulic bender	Electro-hydraulic portable reinforcing bar bender Type: HB-32, 6 each Bending force: Max. 13 ft. Power: 11V/110V 50/60 Hz	Hydraulic bender unit • Hydraulic unit • Hyd. support arm • Arm vertical drive • Arm horizontal drive	200 V 1.5 KW 60 l/min. Five units, span threaded adjust Cylinder stroke 0.38 m Cylinder stroke 0.62 m
Reinforcing bar size	Threaded reinforcing bar D29–D41		
Control method	Sequential control (feedback)	Control method	Push button switch
Size	W 1.3 m × L 12.53 m × H 0.3 m	Size	W 1.75 m × L 8.03 m × H 2.01 m

Sources: Council for Construction Robot Research, 1999, pp. 184–185; Hasegawa (1999).

and heavily distorted threaded reinforcing bars cannot be shaped with this robot. Further specifications of the system are given in Table 2.1.

System 25: Automated Crane for Reinforcing Bar Positioning

Developer: Takenaka Corporation

The automated crane for reinforcing bar positioning from Takenaka was developed to automate and assist with the positioning of reinforcing bars. The

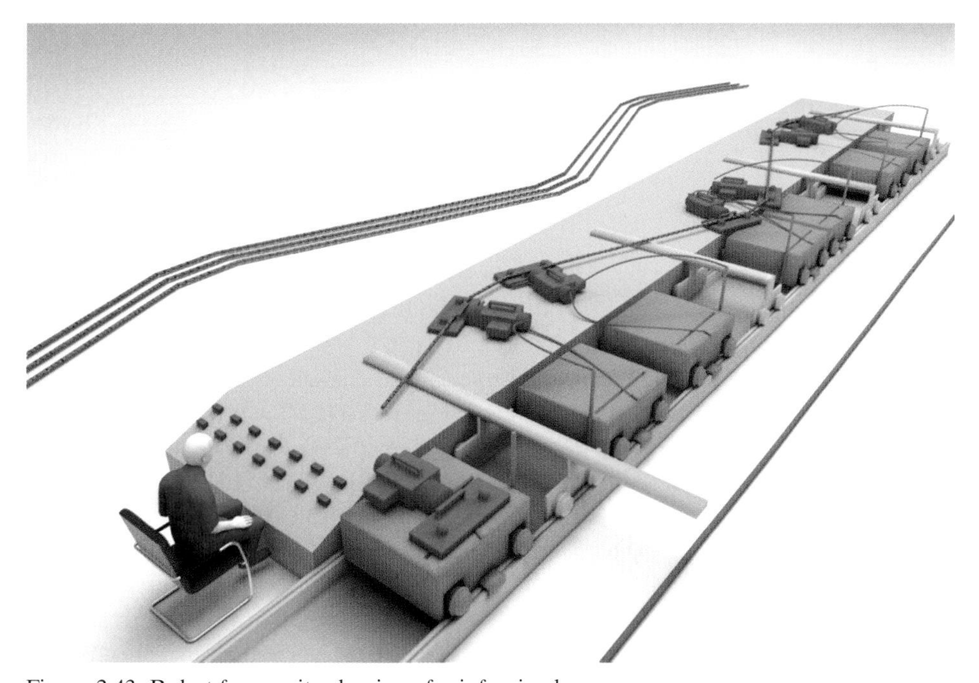

Figure 2.43. Robot for on-site shaping of reinforcing bars.

Figure 2.44. View of automated crane for reinforcing bar positioning system. (Image: Takenaka Corporation.)

loading and placing of the bars can be performed in an automated mode as well as remotely controlled depending on the individual use case. The system to a certain extent builds on a modified crane platform that is equipped with controllable activators, a large telescoping arm, and a specialized end-effector able to effectively manipulate large-diameter reinforcing bars (Figure 2.44). This provides the system with 6 DOF and the ability to perform in confined spaces. Reinforcing bars can be aligned horizontally or vertically and can be placed with an accuracy of 5 centimetres. The control of the crane requires only one human operator. After one bar has been placed (teaching sequence), the system can be set to repeat the same procedure automatically. Installation and deinstallation as well as the work preparation (teaching, programming, etc.) of the system on the site require significant time and resources and it is therefore optimal mainly for large-scale projects. Further specifications of the system are given in Table 2.2.

Table 2.2. *Specifications of the Automated Crane for Reinforcing Bar Positioning System*

Maximum load	150 kg	Motion DOF	6
Lift	30 m	Activator	Induction motor drive
Operating radius	2.5–10 m	Speed control	Inverter system
Weight	10.1 T (crane)	Operation system method	Conventional and wireless remote control
Automated control	Teaching playback	Operating system mode	Manual (lever, voice) or Automatic

Source: Council for Construction Robot Research (1999, pp. 40–41).

System 26: Robot for Positioning of Heavy Reinforcing Bars

Developer: Kajima Corporation

This robot was developed by Kajima to carry out heavy steel reinforcing bar work at construction sites. Structures such as nuclear power plants, dams, and large office or public buildings require huge quantities of reinforced concrete and thus a huge amount of reinforcing bars to be placed on the site. Considering the scale of such projects as well as the size and width of the necessary reinforcing bars, it is physically demanding and monotonous work that can be perfectly substituted by a machine. The robot can carry a set of reinforcing bars and place them in the required sequence. It consists of a tracked mobile platform and on-board storage for carrying a set of steel bars, a manipulator (utilizing relatively simple translational movements only), and an end-effector for handling the steel bars (Figure 2.45). The length of the end-effector is adjustable to be able to cope with a large variety of steel bars of different lengths and hold the reinforcing bars so that they do not bend and can be placed accurately.

Figure 2.45. Robot for positioning of heavy reinforcing bars. (Image: Kajima Corporation.)

System 27: Robot for the Positioning and Joining of Vertical Reinforcing Bars

Developer: Kajima Corporation

When vertical elements such as walls, columns, or pillars are constructed, reinforcing bars need to be added on top of each other and joined prior to moulding and concreting. The reinforcing bars are usually joined by coupling sleeves. In conventional construction, both the positioning of the reinforcing bars as well as the coupling process require human labor and are physically demanding tasks. Kajima's robot for the positioning and joining of vertical reinforcing bars assists the worker in handling and accurately placing the reinforcing bars as well as in the installation/ fixing of the coupling sleeves (Figure 2.46). It consists of a mobile, tracked platform; a manipulator; and an end-effector. The robot can be considered as a cooperative type of robot: the manipulator and the end-effector provide force assistance and are guided intuitively by one worker. The end-effector is able to both hold the reinforcing bar as well to assist with fixing the coupling sleeve.

2.5 Automated/Robotic 3D Concrete Structure Production on the Site

In many regions around the world the basic bearing structure of buildings is made of concrete. Despite the often either vertically or horizontally repetitive concrete structures that constitute the structural basis for concrete buildings, the process of building these structures with *in situ* concrete is highly labour intensive, accounts for a large amount of the overall cost of construction, and builds on either (on-site) custom built or some kind of system formwork. An alternative to this approach provides – at least for high-rise buildings – high-end system formworks (also referred to as automatic or self-climbing formworks) of incumbent providers such Peri or Doka that are equipped with some basic features for automating the climbing and formwork positioning process. Advanced features for such systems (such as additional sensors and activators or advanced digital control and commination capabilities) that would allow those systems almost completely autonomous self-alignment and climbing were the focus of research and development activities in the 1980s and 1990s, but failed to reach the product grade level. Apart from these approaches, recent progress in the field of additive manufacturing/3D printing (and in particular one-stage direct 3D printing) in the non-construction industry has triggered vast interest among researchers and companies to experiment with adapted forms of such approaches in construction. Although this technology is still in an early development stage in the context of construction (in particular in the context of printing speed, printing nozzles/materials, and the orientation of designs of buildings and components towards additive manufacturing significant steps forward would be required), researchers were already able to build prototypes of the equipment and explore a range of alternative methods situated between the deployment of large-sized, gantry-type 3D printers and small-scaled equipment that builds on the cooperation of a swarm of 3D printing robots. Given that research and development get the technology needed under control, the impacts on the flexibility of the construction system, the individuality and functionality of buildings, and construction speed and productivity can be considered tremendous.

Figure 2.46. Robot for positioning and installation of medium-weight reinforcing bars. (Image: Kajima Corporation.)

Automated/Robotic 3D Concrete Structure Production on the Site		
28	Automatic climbing formwork SKE plus	Doka
29	Platform system ACS P	Peri
30	Gantry type contour crafting robot	Behrokh Khoshnevis
31	Cable-suspended robotic contour crafting system	Ohio University
32	Robots equipped with end-effectors for additive manufacturing for extraterrestrial construction	Behrokh Khoshnevis and NASA
33	Minibuilders	Institute for Advance Architecture of Catalonia

System 28: Automatic Climbing Formwork SKE Plus

Developer: Doka

Doka's SKE plus series of automatic climbing formworks is used to produce large reinforced concrete structures such as cores of high-rise buildings, piers, pylons, and towers in a crane-independent manner (Figure 2.47). With its all-hydraulic

Figure 2.47. Automatic climbing formwork SKE50 plus used to build up core and outer structure of a concrete based high-rise building in a crane-independent manner. (Image: © Doka GmbH.)

Figure 2.48. Hydraulic climbing mechanism. (Image: © Doka GmbH.)

equipment (Figures 2.48 and 2.49) large platform gangs, such as all platforms on the outside of a core can be safely raised in a single lifting procedure without open fall hazard locations. The hydraulic lifting process requires two important steps. In the first step the climbing profiles in the climbing shoes anchored on the structure are raised by hydraulic cylinders to the next section. In the second step the climbing scaffolds are pushed upward along the climbing profiles by the same cylinders. This type of climbing formwork is extremely versatile and also allows for climbing inclines, radii, and bends. SKE plus provides multiple working platforms for simultaneous work at several levels and if demanded the possibility of complete enclosure for weather and noise protection.

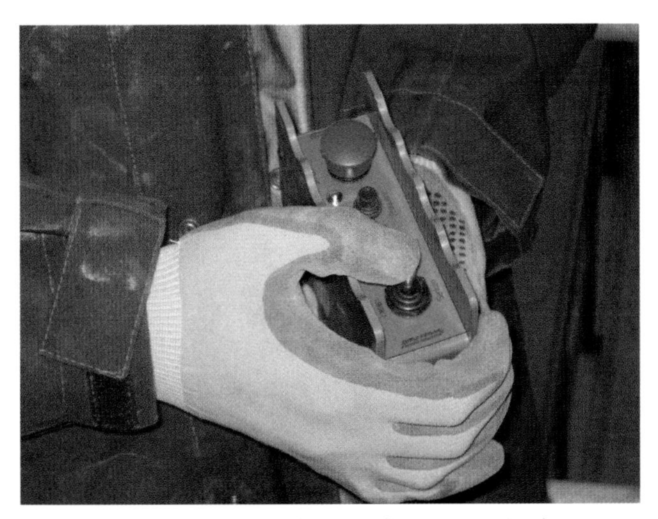

Figure 2.49. Radio control. (Image: © Doka GmbH.)

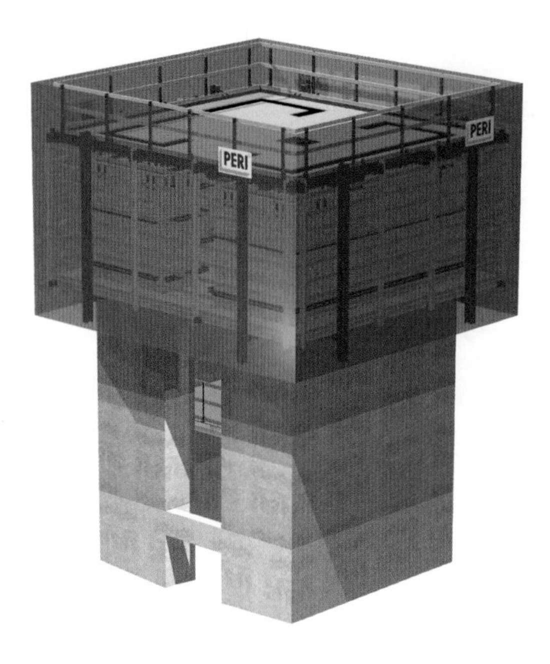

Figure 2.50. Outline of ACS P: The solution for advancing cores of high-rise buildings and tower-like structures. The platforms provide generous storage and working areas. With the ACS P system, only a few platform beams cross the walls. This means that the reinforcement can be partially prefabricated. (Visualization; image: PERI GmbH, Weissenhorn.)

System 29: Platform System ACS P

Developer: Peri

Peri is one of the leading international companies in the formwork sector and, with its platform system ACS P a self-climbing system that allows for crane-independent and partly automated positioning and repositioning of formwork, provides for the on-site production of tower–like structures (Figures 2.50–2.52). As a single entity or divided into various segments the system is pushed upward floor by floor by a hydraulic mechanism. The system provides working platforms on different levels to parallel work activities. The system is strong enough to allow the lifting and storage of material such as the reinforcement. If demanded, a concrete pump can be integrated and move upwards with the system. The ACS is a modular system formwork kit that can be customized by Peri and the clients into a large variety of tower structures.

System 30: Gantry-Type Contour Crafting Robot

Developer: Behrokh Khoshnevis

Behrokh Khoshnevis developed several approaches for additive, layered manufacturing (also referred to as contour crafting) of concrete or concrete-like structures by gantry-type robot systems. The core element of the system is the end-effector (Figures 2.56–2.58) that deposits the material consisting of the nozzle, a material feeding system, and a system of one or several trowels. The trowels play an important role because they help to press the deposited material into the final form and smooth the visible surfaces (Khoshnevis, 2003). Trowel control mechanisms can be used to control the position of the trowels and allow rectangular and also nonrectangular, chamfered edges to be produced. Various types of material depositing end-effectors were developed for the production of simple and also more complex wall sections

Figure 2.51. ACS in action: The three core areas of the DC tower were moved up independently of each other using an ACS climbing formwork up to a height of 220 m. A 3.50-m concreting section was realized by the construction team in 4 days. (Image: PERI GmbH, Weissenhorn.)

Figure 2.52. The PERI climbing mechanism is at the heart of all ACS variants. The positively driven pawl system operates automatically and safely with a lifting speed of 0.5 m/minute. (Image: PERI GmbH, Weissenhorn.)

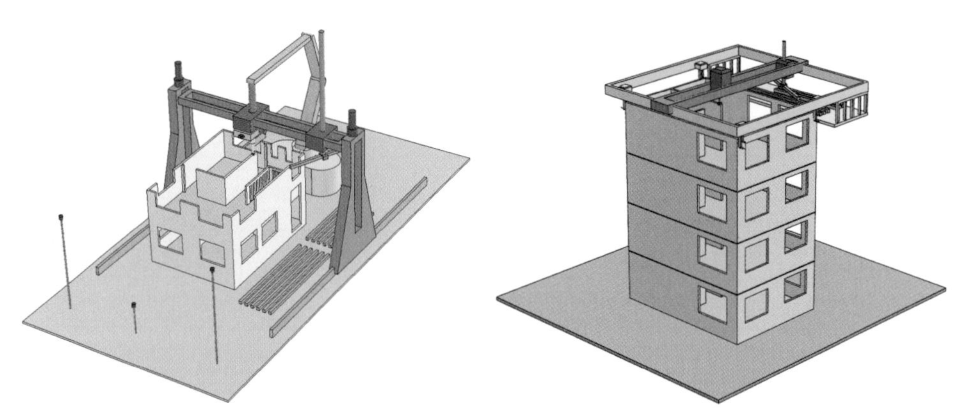

Figure 2.53. Gantry-type system for smaller buildings. (Image: Behrokh Khoshnevis, Center for Rapid Automated Fabrication Technologies [CRAFT], University of Southern California.)

Figure 2.54. Gantry-type system for vertically oriented buildings. (Image: Behrokh Khoshnevis, Center for Rapid Automated Fabrication Technologies [CRAFT, University of Southern California.)

and structures. Khoshnevis also developed approaches for the insertion of reinforcement by (1) producing walls with a hollow space in the middle that is subsequently filled with reinforcement and concrete; (2) stacking small reinforcement elements on each other layer by layer and overprinting them layer by layer, or by using material containing crushed fibres. Furthermore, they developed approaches for the insertion of elements into the wall structure produced that contain the plumbing and electricity/communication infrastructure. The gantry-type system can be deployed on the construction site in various configurations (Figures 2.53–2.55) depending

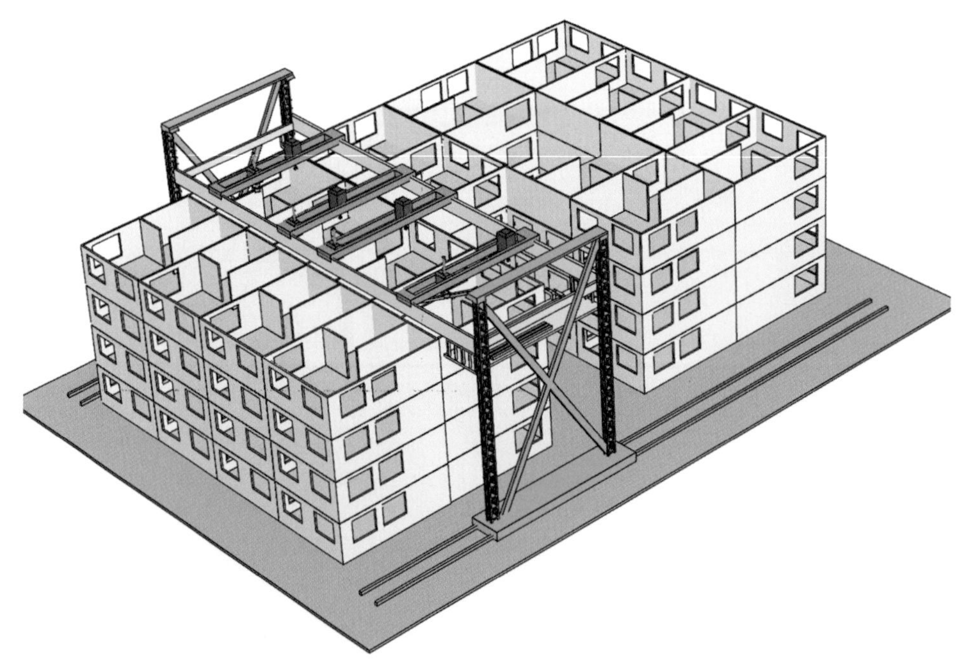

Figure 2.55. Gantry-type system for larger buildings/condominiums. (Image: Behrokh Khoshnevis, Center for Rapid Automated Fabrication Technologies [CRAFT], University of Southern California.)

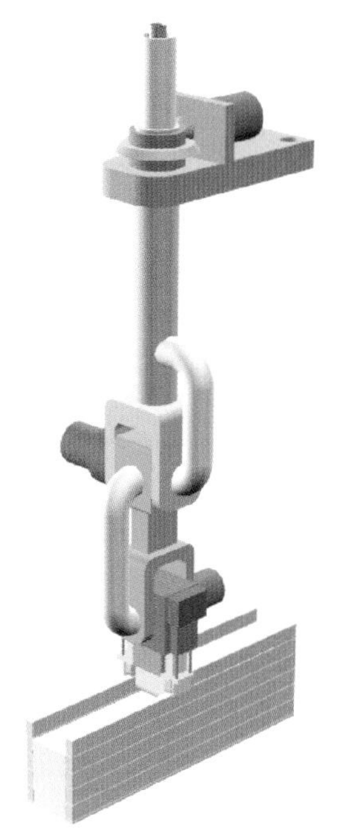

Figure 2.56. Example of the composition of an end-effector. (Image: Behrokh Khoshnevis, Center for Rapid Automated Fabrication Technologies [CRAFT, University of Southern California.)

on the type of building to be constructed and it can in addition serve as the end-effector for material deposition and be equipped with additional manipulators or end-effectors for reinforcement placement, for example. Recent research at Khoshnevis has focussed on the optimization of the material deposition process in multinozzle, multigantry systems (Zhang & Khoshnevis 2012.)

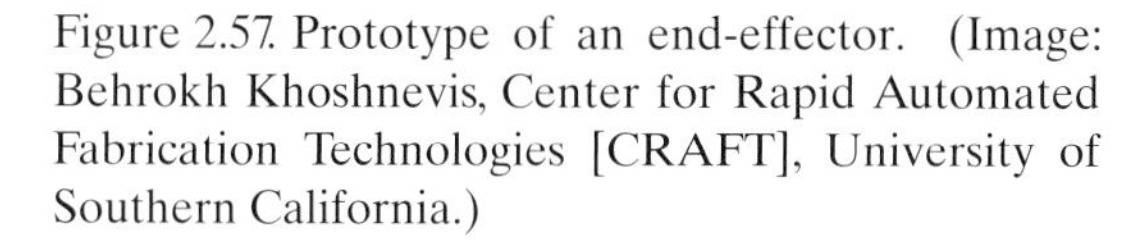

Figure 2.57. Prototype of an end-effector. (Image: Behrokh Khoshnevis, Center for Rapid Automated Fabrication Technologies [CRAFT], University of Southern California.)

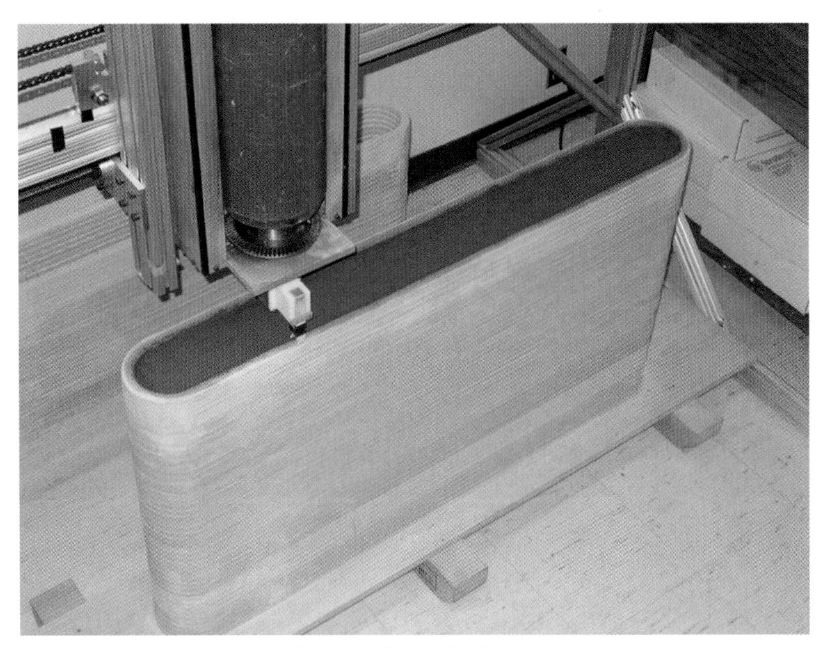

Figure 2.58. Prototypic production of a wall. (Image: Behrokh Khoshnevis, Center for Rapid Automated Fabrication Technologies [CRAFT], University of Southern California.)

System 31: Cable-Suspended Robotic Contour Crafting System

Developer: Ohio University, Department of Mechanical Engineering and Department of Civil Engineering

Bosscher et al. (2007) suggest as an alternative to gantry-type contour crafting systems a cable-suspended robotic contour crafting system (Cartesian cable robot; Figure 2.59). Contour crafting requires that the end-effector with the nozzle for material deposition is moved through relatively large work space depending on the size of the building to be built. A cable-suspended system can cover large workspaces

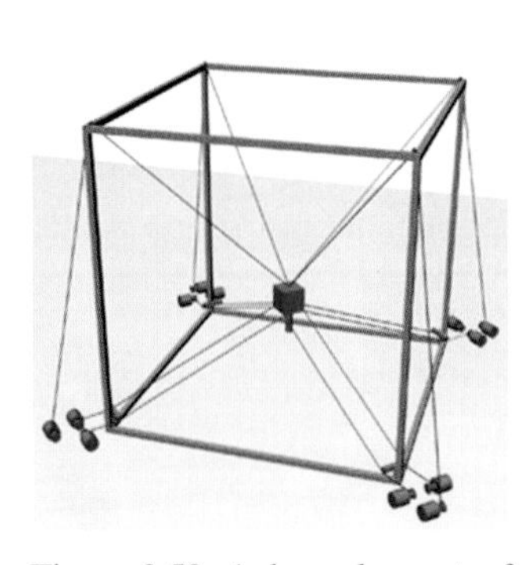

Figure 2.59. A key element of the contour crafting system is a Cartesian cable robot (Reprinted from Bosscher and Williams, 2007, "Cable-suspended robotic contour crafting system," *Automation in Construction*, 17:45–55, with permission from Elsevier.)

Figure 2.60. Crossbar of the cable robot in lowered (left) and lifted (right) mode (Reprinted from Bosscher and Williams, 2007, "Cable-suspended robotic contour crafting system," *Automation in Construction*, 17:45–55, with permission from Elsevier.)

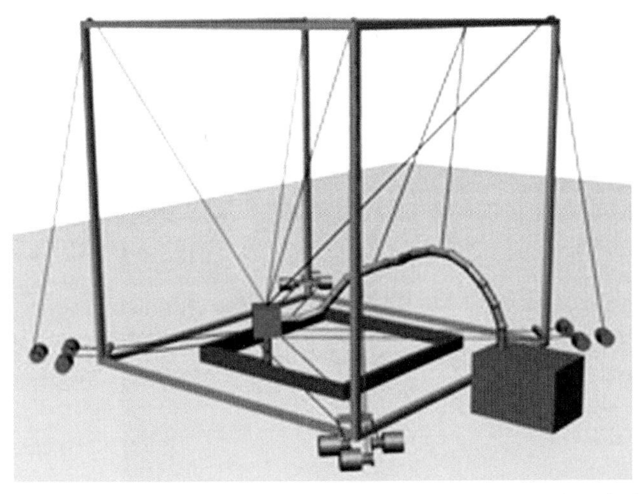

Figure 2.61. The system building a structure with crossbars being lifted step by step (Reprinted from Bosscher and Williams, 2007, "Cable-suspended robotic contour crafting system," *Automation in Construction*, 17:45–55, with permission from Elsevier.)

and is in contrast to other solutions relatively easily deployable on the site. The proposed contour crafting system consists of a frame and a suspended end-effector held and operated by eight lower and four upper cables. The robot system can be operated by translation-only motion (upper cables carry the weight and the lower cables perform the translation-only motion), significantly simplifying kinematics and operation of the system. The system is able to position and orient the end-effector in the operational layer or plane. To avoid collision between the lower cables and the structures to be produced in the frame, crossbars allow the lower cables to be lifted accordingly (Figures 2.60 and 2.61). Bosscher et al. (2007) analysed the kinematic structure, the robot statics, robot workspace characteristics, and cost and productivity of the cable-suspended robotic contour crafting system to determine its feasibility.

System 32: Robots Equipped with End-Effectors for Additive Manufacturing for Extraterrestrial Construction

Developer: Center for Rapid Automated Fabrication Technologies (Behrokh Khoshnevis) in cooperation with NASA

This robot system was developed by Prof. Khoshnevis and NASA as part of a NASA Innovative Advanced Concept (NIAC) project. The project builds partly on NASA's development of novel equipment (rovers, space suits, etc.). The core idea of the concept is to equip robotic rovers (which are derived from earlier NASA concepts for robotic construction rovers) with end-effectors for the additive manufacturing of structures such as landing pads and the basic structure of shelters in extraterrestrial environments (Figure 2.62). The approach would allow utilization of material to be found in the environment (also referred to as *in situ* resource utilization), transform it into a concrete-like substance, and feed it over the robots into the "printing" end-effectors. It is envisioned that multiple robots would work in cooperation with other robotic construction machines and human beings. Among the components of

Figure 2.62. Multiple robots for additive manufacturing work in cooperation with other robotic construction machines and human beings. (Image: Behrokh Khoshnevis, Center for Rapid Automated Fabrication Technologies [CRAFT], University of Southern California; Rendering courtesy of Behnaz Farahi and Connor Wingfield.)

the project were a feasibility study, the development of mock-ups, and the testing and real-world simulation with some existing equipment in NASA test facilities on earth.

System 33: Minibuilders

Developer: Institute for Advanced Architecture of Catalonia (IAAC)

One of the biggest weaknesses of 3D printing that represents a challenge for its efficient use on construction sites is the usually low printing speed and the often large scale of the structures that have to be built. Minibuilders therefore uses a swarm approach in which many small and relatively simple robots with dedicated, assigned tasks cooperate. The robots can work in parallel, which increases speed, and the system is not dependent on the size and shape of the building as, for example, in contour crafting. The building material can be brought centrally or decentrally from (if necessary also mobile) material supply robot platforms (Figures 2.63 and 2.64). The approach distinguishes among three categories of printing robots (Figures 2.65–2.68): (1) foundation printing robots (able to move on the ground; specialized on printing the first layers and providing a stable foundation structure); (2) grip robots (able to move on the foundation structure; specialized in printing and extending the structure); and (3) shell vacuum robots (able to move on the outside of the printed structures and add further layers for reinforcement and finishing). All Minibuilder robots follow a modular approach, and a frame made of profiles can, depending on factors as building, type, dimensions, building material, be equipped with the necessary locomotion and printing tools.

For further information, see IAAC (2015).

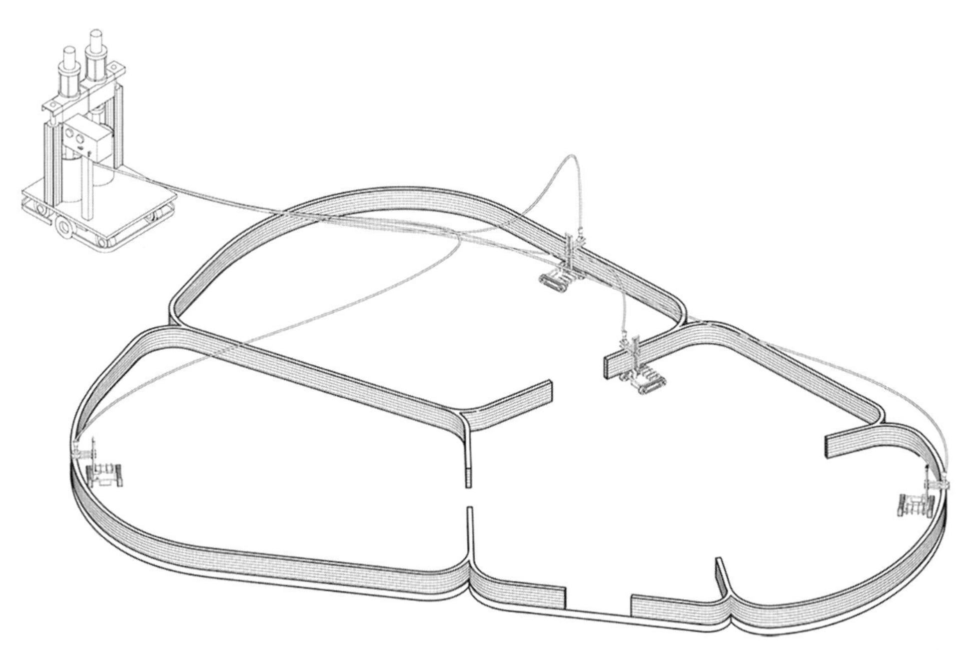

Figure 2.63. Overall concept; initial construction phase (foundation printing robots at work). (Image: Institute for Advance Architecture of Catalonia.)

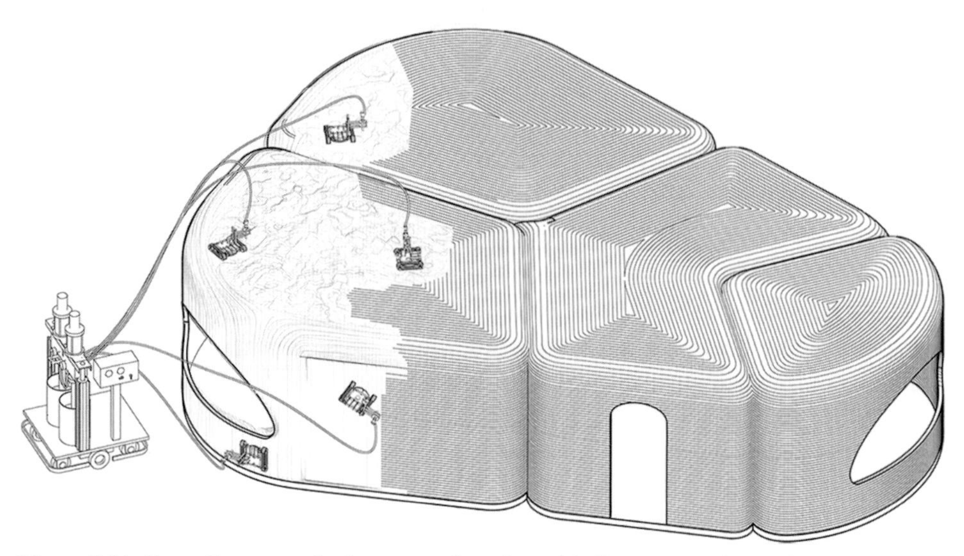

Figure 2.64. Overall concept, final construction phase (shell vacuum robots adding additional layers for reinforcement). (Image: Institute for Advance Architecture of Catalonia.)

2.6 Automated/Robotic 3D Truss/Steel Structure Assembly on the Site

Steel-based structures, truss structures, and kits used in the context of building and habitat construction on earth (such as halls, hangars, factories, large convention centres, sport centres, train station airports, etc.), in extreme environments on earth (Arctic construction, underwater construction, etc.), as well as in space (space stations, habitat on moon/mars, etc.) are usually based on the restitutive assembly of a limited amount of elements such as links, nodes, and panels. Standardization and repetition, which are inherent elements of the composition and assembly of such elements, make automation a viable option. However, the issue is tricky because these steel structures require both complex joining systems and joining operations as well as a high level of dexterity and accuracy of the assembly machines. Furthermore, in reality structures and individual components might be large-scale and heavy in weight, which makes it difficult to handle and join individual elements accurately and

Figure 2.65. Testing foundation printing robots. (Image: Institute for Advance Architecture of Catalonia.)

Figure 2.66. Grip robots for printing of the main structure. (Image: Institute for Advance Architecture of Catalonia.)

Figure 2.67. Detail of grip robot. (Image: Institute for Advance Architecture of Catalonia.)

safely. As a result of these obstacles, systems in this category represent only early lab and feasibility demonstration projects. Nevertheless, recent advances in robotics also promise advances in the field of robotic steel structure assembly, and new technologies such as 3D printing also allow for new approaches and designs with respect to steel links, nodes, and joining systems. The examples outlined in the following indicate that for efficient assembly an extremely deep integration and coadaptation of the building system, nodes, steel links, kinematic structure of the robots, and end-effectors is required.

Automated/Robotic 3D Truss/Steel Structure Assembly on the Site		
34	Location orientation manipulator (LOM)	Konrad Wachsmann
35	System for robotic assembly of truss structures	NASA Langley Research Center

Figure 2.68. Shell vacuum robots adding additional layers for reinforcement. (Image: Institute for Advance Architecture of Catalonia.)

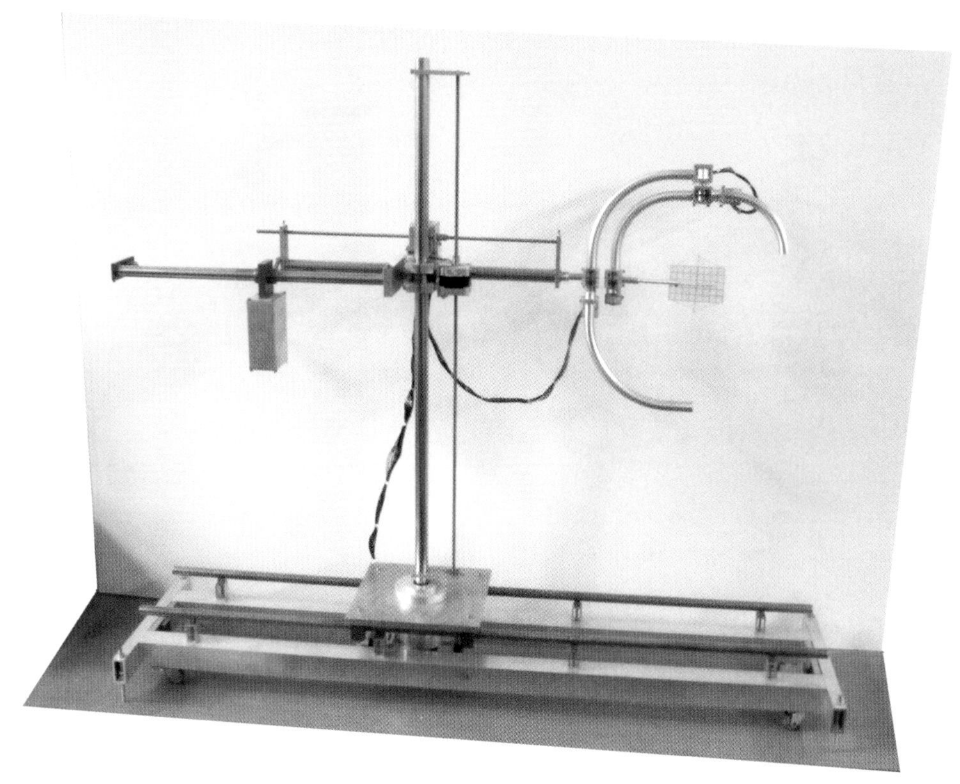

Figure 2.69-1. LOM 7-DOF prototype manipulator for on-site steel system assembly. (Scale model: Part of a master's thesis by W. Lauer supervised by Prof. T. Bock and Prof. W. Nerdinger. Image: Chair of Building Realization and Robotics.)

System 34: Location-Orientation Manipulator (LOM)

Developer: Konrad Wachsmann

In 1969, an early prototype for an on-site manipulator for steel structure assembly (Figure 2.69-1) was developed by Konrad Wachsmann and his students Bollinger and Mendoza. Wachsmann was an engineer who developed several kit systems (e.g., mobilar structure, USAF hangar structure, grapevine structure; see, e.g., Wachsmann 1969; Nerdinger 2010) that allowed the assembly of truss-based buildings such as halls, factories, and hangars on the construction site from sets of standardized and prefabricated basic components (representing links and joints; see, e.g., Figure 2.69-2). It is assumed that the LOM was developed by Wachsmann as a system that would be able to position and align these prefabricated basic components in the required repetitive manner in a 3D work space on the construction site (Lauer & Bock 2010). The complex kinematic structure of the LOM in combination with 7 DOF indicates that the "robot" was intended to execute a variety of highly complex assembly operations (which are necessary for the assembly of the steel structures in question). Drawings and sketches of Konrad Wachsmann indicate that it was planned to let multiple LOMs cooperatively and in parallel execute the positing, alignment, and fixation of basic components (Lauer and Bock 2010) and integrate them with

Figure 2.69-2. Example of a structure that Konrad Wachsmann might have had in mind for processing by LOMs; the image shows a replica model of a USAF hangar structure planned by Wachsmann. (Image: © Architekturmuseum der TU München, Signatur wachs.-3-1; modell making: Florian Schorer, Quirin Stoiber, built 2009.)

other manufacturing technology to larger, flexible on-site production systems (Bock et al. 2010).

System 35: System for Robotic Assembly of Truss Structures
Developer: NASA Langley Research Center, W. Doggett

This system was developed by NASA Langley Research Center as a prototype for the assembly of space structures that are assumed to consist of a multitude of members of equal length. A laboratory test bed was set up (see also Doggett 2002) that consisted of a gantry-type motion base by which a robot arm with an end-effector and pallets, which provides the construction elements (truss struts, etc.) to the robot, is moved through/along the work area as well as a rotational motion base that is able to rotate and reposition the truss structure (Figure 2.70). Basic elements of the systemized truss structure such as the nodes and truss strut elements were geometrically adapted, in terms of alignment and connection mechanisms, to the robotic assembly process. The node, for example, is equipped with machine vision targets and dedicated alignment grooves (Figure 2.71). An end-effector was developed that can insert the truss struts into the alignment grooves of the nodes and activate the locking nut of the connection mechanism (Figure 2.72). The end-effector of the robot could be exchanged with a second end-effector that was developed for the handling and installation of panels in the structures.

2.7 Bricklaying Robots

Despite advances in the prefabrication of brick walls and in the use and processing of other key building materials such as concrete, steel, and wood, the on-site production

Figure 2.70. Laboratory test bed for the assembly of complex truss structures. (Image: Doggett, William / NASA Langley Research Center.)

Figure 2.71. Basic elements of the truss structure were fully adapted to the robotic assembly process. The nodes, for example, are equipped with machine vision targets and dedicated alignment grooves. (Image: Doggett, William/NASA Langley Research Center.)

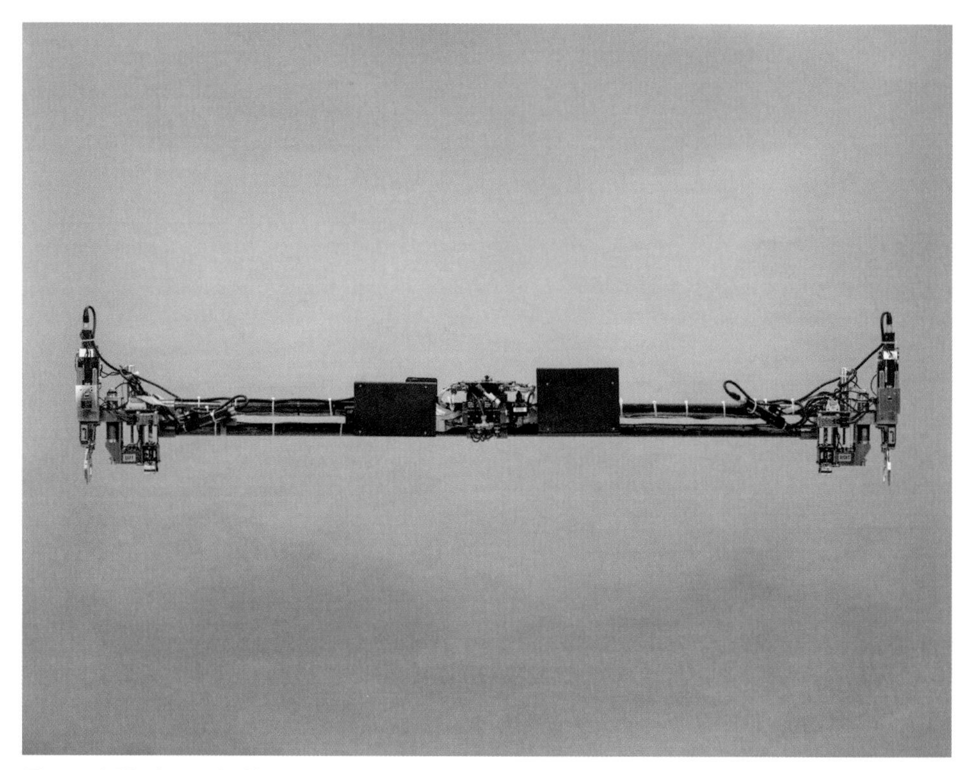

Figure 2.72. An end-effector was developed that is able to insert the truss struts into the alignment grooves of the nodes and activate the locking nut of the connection mechanism. (Image: Doggett, William/NASA Langley Research Center.)

of brick structures continues to be highly important. In particular, in the context of housing and smaller public and office buildings, brickwork is highly regarded in many countries by both investors and customers/inhabitants because of its positive impact on the living comfort and the (climatic) conditions it is able to create in buildings and thus the value it adds to buildings. Apart from that, the long history of brickwork use and research and development in the field have made bricks an advanced building material that is today available in many different forms and even as high-tech versions enriched, for example, with integrated and highly advanced insulation features. The aforementioned advantages and advances along with the capability of brickwork producers nowadays to customize and sequence bricks and their delivery to individual customers on demand and just in time have led to a revival of the approach of using robots to create brickwork structures on the site. Whereas the approach was highly popular in the early days of the STCR era (SMAS, BRONCO, ROCCO), it almost vanished for a decade or two in light of other approaches and developments. However, very recently research and development in the field was again intensified, and apart from new conceptual approaches (Mike Silver, Mateus Lemes de Aguiar), two companies (Fastbrick Robotics and Construction Robotics) actually succeeded in introducing robust and marketable versions of bricklaying robots.

Bricklaying Robots		
36	On-site brickwork laying system: solid material assembly system (SMAS)	Japanese Research Institute, Prof. Kodama
37	On-site brickwork laying robot (BRONCO)	Institute of Control Technology for Machine Tools and Manufacturing Systems at the University of Stuttgart
38	Robotic construction system for computer-integrated construction (ROCCO)	Lissmac and Karlsruhe University
39	On-site construction robots (OSCR) series for masonry construction	Mike Silver
40	Bricklaying robot concept	University of Toronto/Mateus Lemes de Aguiar
41	Hadrian bricklaying robot	Fastbrick Robotics
42	Semi-Automatic masonry system (SAM)	Construction Robotics

System 36: On-Site Brickwork Laying System: Solid Material Assembly System (SMAS)

Developer: Japanese Research Institute

The solid material assembly system (SMAS) was developed by the Japanese Research Institute for the automation of the on-site construction of structural and nonstructural walls on construction sites. A reinforced concrete wall system was redesigned in terms of structure, component design, and variation control to allow for efficient robotic assembly. The robot-adapted building blocks were hollow on the inside (so they were lightweight and thus could be optimally handled by the robot) and the edges where chamfered according to robot-oriented design guidelines to compensate positioning inaccuracies and guide the building blocks into place. The robot system positions, aligns, and fixes building blocks and an integrated reinforcement (Figure 2.73). The system consists of a mobile base, a manipulator, a component adapted end-effector (Figure 2.74), a pallet-based material delivery system providing the building blocks in a sequenced and standardized way to the end-effector (Figure 2.75), and last but not least building blocks that were considered as part of the system (and therefore were fully adapted to the end-effector and the automated installation process; Figure 2.76). The robot moves to the pallet, picks up a building block, transports it to the designated stacking position, and stacks it. The picking up of building blocks from the pallets in order is based on a depalletizing program that was originally developed for general use in the manufacturing industry. After the robot releases the building blocks on the designated position, it screws also a bolt serving as a vertical reinforcing bar into a nut on the underlaid building block and thus fixes the building block. The reinforcing bar also serves as a guiding pin for the end-effector during component alignment. Once the wall is built up, the erected structure serves as a kind of moulding and is filled with concrete.

A more detailed description of the individual subsystems is given in **Volume 1**.

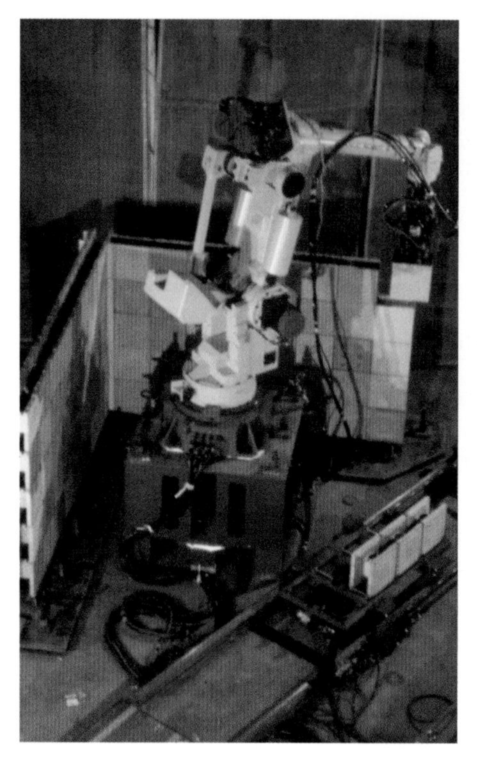

Figure 2.73. View of total SMAS system in action.

Figure 2.74. End-effector gripping a component.

Figure 2.75. Pallet for handling and presenting eight elements on standardized slots to the robot.

Figure 2.76. Building components designed according to robot-oriented design guidelines delivered on the pallet.

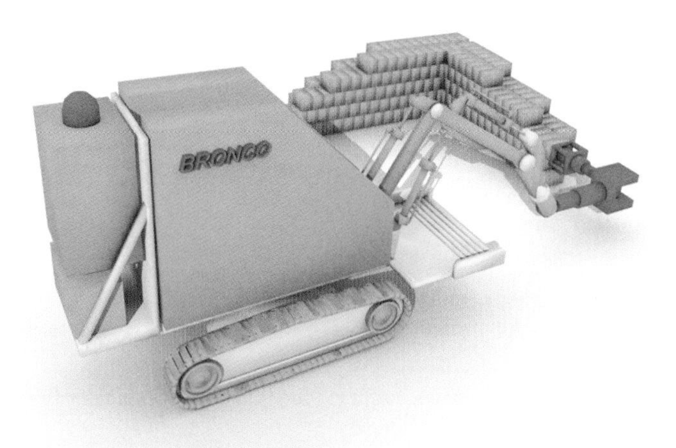

Figure 2.77. On-site brickwork laying robot BRONCO. (Drawing according to Pritschow et al. 1998.)

System 37: On-Site Brickwork-Laying Robot BRONCO

Developer: Institute of Control Technology for Machine Tools and Manufacturing Systems at the University of Stuttgart

Developed by the Institute of Control Technology for Machine Tools and Manufacturing Systems at the University of Stuttgart, BRONCO is used for the automation of bricklaying in residential construction (Figures 2.77 and 2.78). The robot system consists of a tracked mobile base, a 7-DOF manipulator installed on top of this base, an end-effector designed for gripping bricks, and a platform attached to the mobile base (multifunctional technological unit) that serves as the on-board mortar application system. The end-effector is equipped with a vacuum suction system as

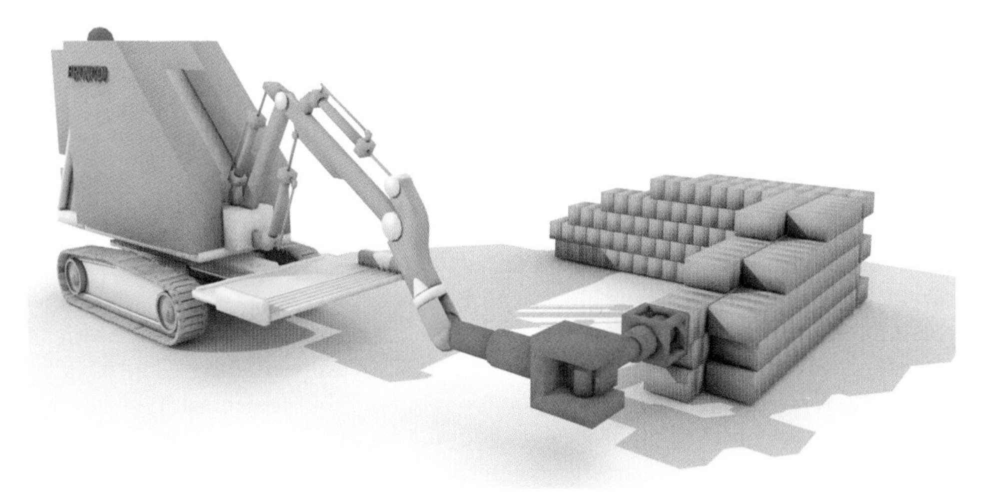

Figure 2.78. 3D model of BRONCO robot fixing brickwork. (Drawing according to Pritschow et al. 1998.)

well as distance and inclination sensors. The robot takes the brick from standard pallets, places it on the multifunctional technological unit where the brick is measured, mortar is applied to the brick, and the brick finally is centred and presented to the robot in a predefined way (presenting the brick correctly to the end-effector with respect to the tool centre point is a prerequisite for accurate positioning, alignment, and fixation). Whereas the bricks are taken up from the pallet and placed on the multifunctional technological unit by the end-effector grasping the brick from atop, the actual assembly is conducted by the end-effector grasping the brick from the side to further support accurate positioning, alignment, and fixation. A brickwork planning system generates the manufacturing outline (necessary motions, etc.) from computer-aided design (CAD) data automatically. The robot system can be remotely supervised and/or directly remotely controlled through a handheld-type control panel.

For further information, see Pritschow et al. (1995, 1998) and Dalacker (1997).

System 38: Robotic Construction System for Computer-Integrated Construction (ROCCO)

Developer: Lissmac and Karlsruhe University

ROCCO is a robot construction system for computer-integrated construction. Under the project name ROCCO, the Lissmac and Karlsruhe consortium developed within the framework of an Esprit (EU ESPRIT 3 6450 project) research project a system that enables fully automated masonry construction on-site. Companies and institutes participated in the EU-funded project in an interdisciplinary and international approach with experts in the fields of construction technology, mechanical and electrical engineering, and information technology from Germany, Spain, and Belgium. The goal of the project was the development of a computer-integrated robot system (Figure 2.79) that also contains an integrated information and communications technology solution for all steps from the architectural design to the automated assembly of the components on the construction site. The main emphasis of the project was on the realization of a mobile robot system for construction site operation as well as on the integration of a computer-based system for working preparation and quality control. In the working preparation phase, the necessary data are generated for the prefabrication and customization of the masonry blocks, for the construction site layout, and for the automatic robot program generation.

Based on the CAD representation of the building, first the walls are divided automatically into the necessary blocks. In the next step the optimal working positions of the mobile robot are calculated automatically as well as the positions of the pallets and the arrangement of the blocks on the pallets. With this information, the necessary nonstandard and standard blocks can be produced and palletized. Finally, the robot motions are generated automatically from the calculated geometry information. The user interface on the construction site is graphically interactive and enables the user to reprogram partially the generated robot motions to deal with the imponderability of the construction process without the necessity to learn a specific robot programming language. In the context of the research project, several automated and less automated variations of the system were generated and tested. Furthermore, a variety of end-effectors were developed and tested (Figures 2.80 and 2.81).

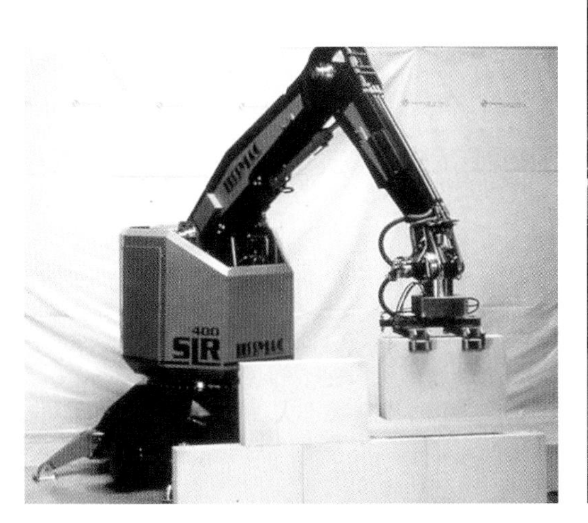

Figure 2.79. On-site brickwork laying robot ROCCO.

Figure 2.80. Vacuum suction mechanism based end-effector developed in the research project ROCCO.

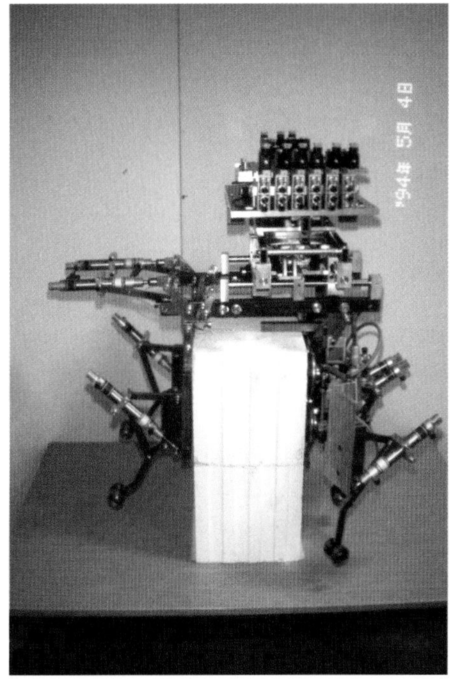

Figure 2.81. Embracing type end-effector developed in the research project ROCCO.

Figure 2.82. Envisioned use case for the operation of OSCR 4. (Image: Mike Silver.)

For further information, see Andres et al. (1994), Bock and Gebhart (1994), and Bock and Steffani (1994).

System 39: On-Site Construction Robot (OSCR) Series for Masonry Construction

Developers: Mike Silver; Construction Site Robotics Lab (CSRL), Ball State University; and Rust Belt Robotics Lab, University of Buffalo

The OSCR series for masonry construction represents a novel approach towards automating or robotizing construction with bricks and stones (Figures 2.82–2.86). Unlike other examples mentioned in this book in which robot arms are put on mobile platforms or lifters and completely automate the arrangement of the stones while at the same time – for example, owing to safety reasons – excluding human beings from the area where they are operated, in this approach simple bi- and quadrupedal robots are suggested that can be deployed on the site as swarms of robots. The OSCR series robots shall be able to interact with human beings in the dynamic environment

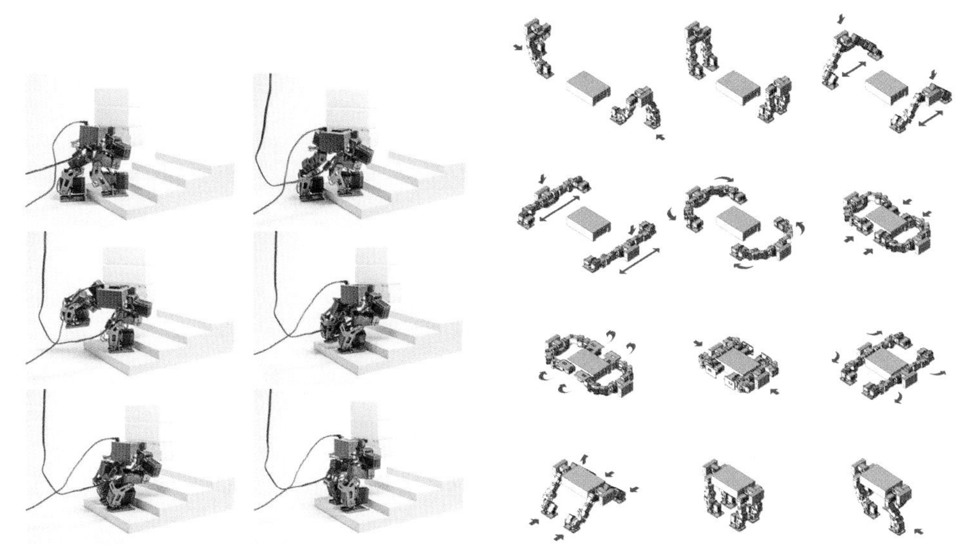

Figure 2.83. OSCR 2 test. (Image: Paul Qaysi.)

Figure 2.84. Possible motions and configurations of OSCR 2. (Image: Mike Silver.)

of conventional or partly conventional construction sites. Furthermore, OSCRs can individually handle stones or couple (e.g., two OSCR 2) and jointly handle stones depending on the payloads and tasks to be accomplished. So far, four OSCR series models were developed and their functional mock-ups were built and tested. The OSCR series robots build on a strictly modular hardware approach and advanced vision systems. Each robot was developed by thoroughly analysing work processes in cooperation with construction workers and by transferring the outcomes of these analyses into distinct robot configurations.

For further information, see Silver (2014).

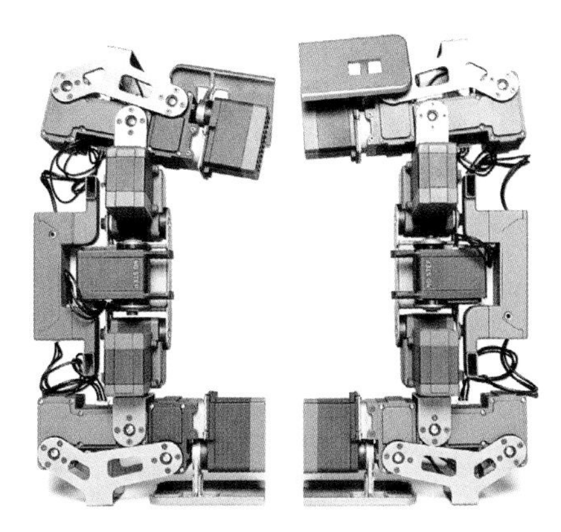

Figure 2.85. OSCR 2 mock-up. (Image: Paul Qaysi.)

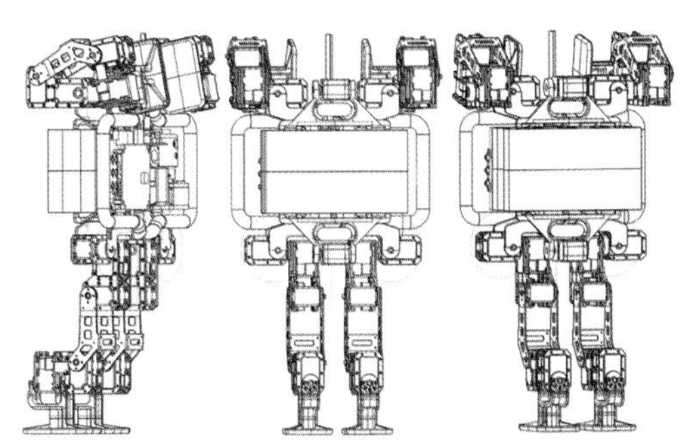

Figure 2.86. DWG drawings of OSCR 4. (Image: Mike Silver.)

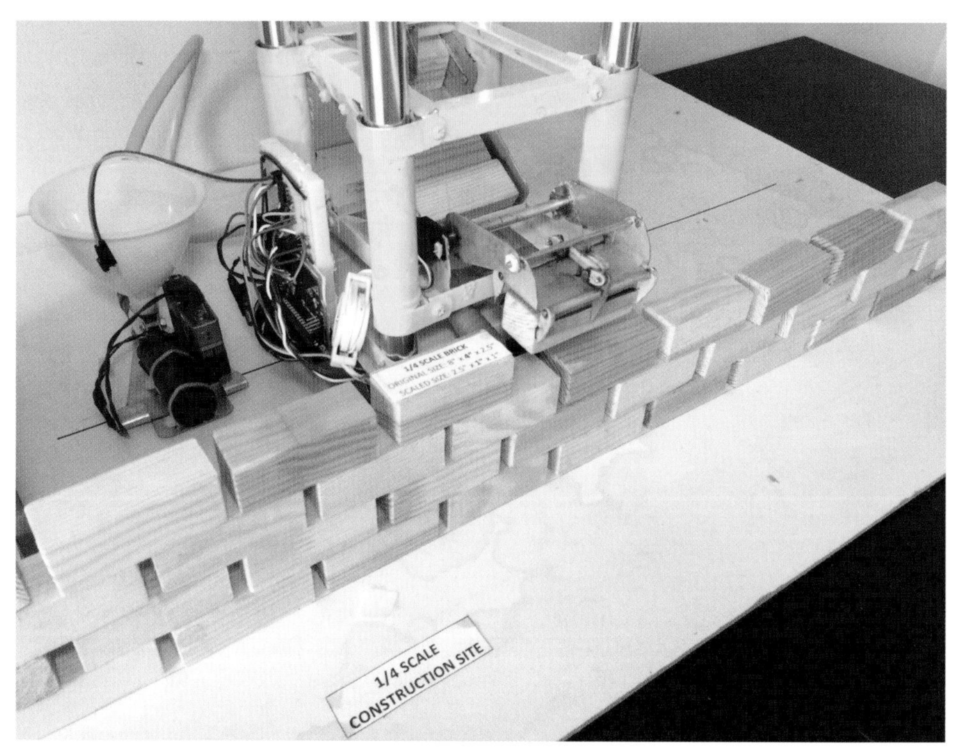

Figure 2.87. Mock-up of bricklaying robot. (Image: Mateus L de Aguiar.)

System 40: Bricklaying Robot Concept

Developer: Mateus Lemes de Aguiar; Department of Mechanical and Industrial Engineering, University of Toronto; Supervisor: Prof. Dr. Kamran Behdinan; grant institution: Ciências Sem Fronteiras – CNPq.

The bricklaying robot (Figures 2.87–2.89) was developed by Mateus Lemes de Aguiar as part of his master's thesis at the Department of Mechanical and Industrial Engineering, University of Toronto. The robot follows a simple and repetitive

Figure 2.88. Visualization of the bricklaying robot system. (Image: Mateus L de Aguiar.)

Figure 2.89. Mounted to the frame are, on one side, a mortar dispenser and, on the other side, an alignment mechanism. (Image: Mateus L de Aguiar.)

approach. The robot can move only in one direction along the wall to be built. Bricks are fed by the human operator or an additional feeding system into a flow rack. A robot integrated into a base frame set on top of a mobile platform picks the bricks up one by one by a simple back and forth movement. Mounted to the frame are, on one side (the side of forward movement), a mortar dispenser (putting mortar in a continuous line on the bricks of the lower, already constructed brick level) and, on the other side, an alignment mechanism including a sensor system for accuracy measurement.

System 41: Hadrian Bricklaying Robot

Developer: Fastbrick Robotics

The Australian start-up company Fastbrick Robotics developed 3D automated robotic bricklaying technology and the Hadrian bricklaying robot for the automated production of brick-based buildings (Figure 2.90). Standard brick pallets are placed in a magazine on the machine (Figure 2.96) where three gantry-type manipulators feed the bricks to a chain-based logistics system (Figure 2.94). This logistics system transports the bricks in a continuous stream over the boom of the machine (Figure 2.95) to the end-effector placing the bricks. The end-effector is the system's key element (Figures 2.91–2.93). A mechanism applies concrete to the bricks before they

Figure 2.90. Overview of Hadrian bricklaying robot system; a laser guidance system allows accurate placement of the bricks. (Visualization; image: Fastbrick Robotics.)

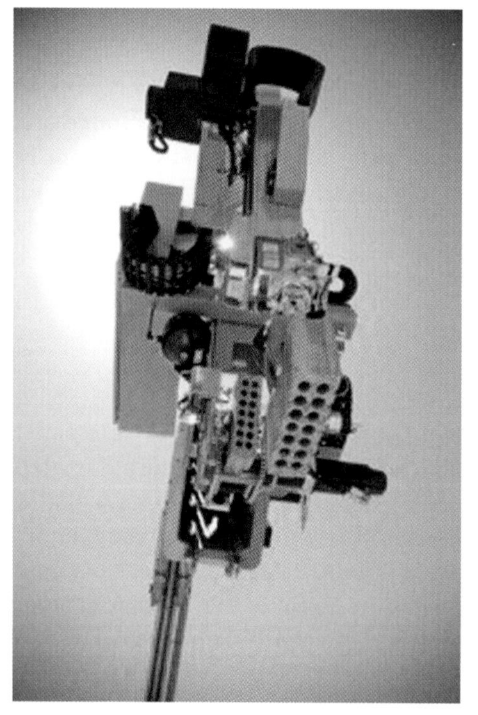

Figure 2.91. End-effector, front view. (Visualization; image: Fastbrick Robotics.)

Figure 2.92. End-effector, detail. (Visualization; image: Fastbrick Robotics.)

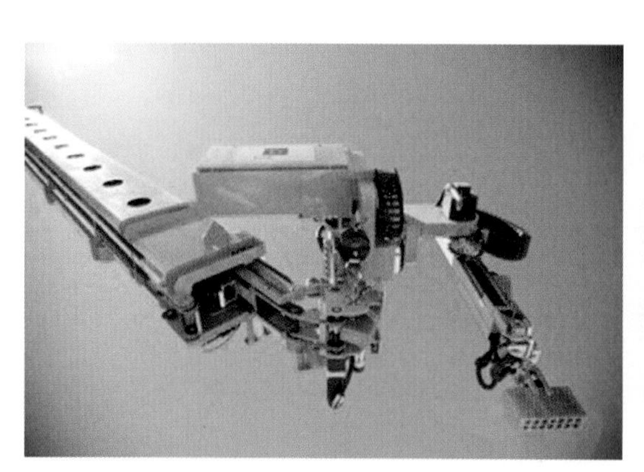

Figure 2.93. End-effector, side view. (Visualization; image: Fastbrick Robotics.)

Figure 2.94. Picking up of bricks from the magazine on the robot. (Visualization; image: Fastbrick Robotics.)

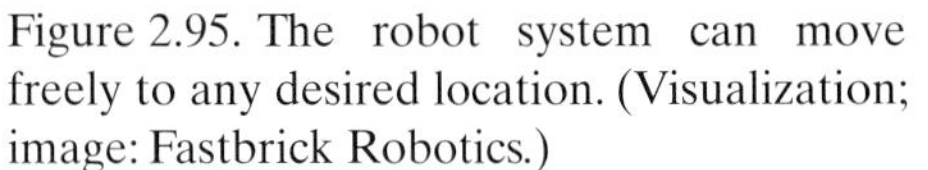

Figure 2.95. The robot system can move freely to any desired location. (Visualization; image: Fastbrick Robotics.)

Figure 2.96. Brick magazine on the robot. (Visualization; image: Fastbrick Robotics.)

are released from the chain-based logistics system and presents them to a manipulator that conducts final positioning and fixation. A laser guidance system allows for autonomous navigation and positioning and ensures a super-accurate placement (accuracy of brick positioning of ±0.5 mm).

For further information, see Fastbrick Robotics (2015).

System 42: Semi-Automatic Masonry System (SAM)

Developer: Construction Robotics

The semi-automated masonry system (SAM) is a robotic system developed by the American company Construction Robotics. The robot is able to perform the repetitive task of brick laying in an efficient, safe, accurate, and predictable way. The system consists of (1) a compact, mobile, main module (contains material feeding systems for mortar and bricks, the robot, security systems that stop the robot when collisions with scaffolding or human operators are imminent, a mortar dispensing device, and a mobile platform for horizontal movement; Figures 2.97–2.99); a (2) lifting platform (lifting main module and human operators; Figure 2.100); and (3) a control and

Figure 2.97. Compact, mobile, main SAM module. (Image: Construction Robotics.)

Figure 2.98. Moving SAM at the World of Concrete with the XTREME lift – SAM main module can easily be handled on the site. (Image: Construction Robotics.)

Figure 2.99. The main SAM module is positioned on the lifting platform. (Image: Construction Robotics.)

Figure 2.100. The main SAM module operates horizontally along the lifting platform and is moved vertically by the lifting platform. (Image: Construction Robotics.)

software system. On the site the main module is placed on the lifting platform, where it can drive along in a horizontal direction and build a certain number of brick layers. After that the platform is lifted and the process is repeated. A variety of brick types and sizes can be processed. The bricks are manually loaded into conveyor-like feeding system. The bricks are then individually picked up by the robot (Figure 2.101) and oriented to a mortar dispensing device (Figure 2.102), where mortar is added to one or two of the brick's sides. With the help of a laser guided sensing system, the end-effector (Figures 2.103 and 2.104) places the bricks with a high degree of accuracy. As the robot moves along the lifting platform, it is up to human workers to remove excess mortar from the wall before it dries up. The newly released version of the system (SAM100) is smaller in size, lighter, and almost three times faster

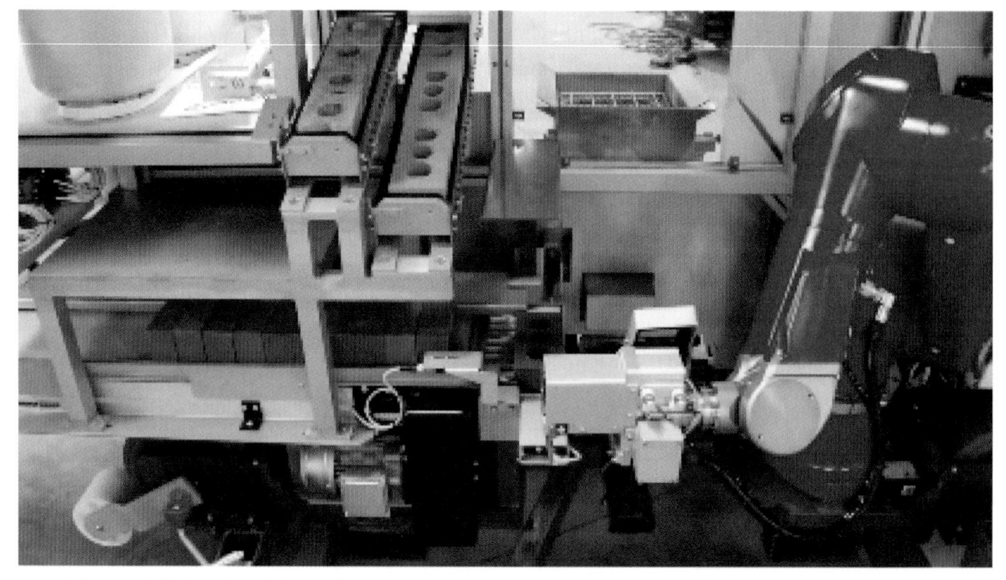
Figure 2.101. Material feeding system for SAM. (Image: Construction Robotics.)

Figure 2.102. Mortar application through cooperation of robot and mortar dispensing device. (Image: Construction Robotics.)

Figure 2.103. The end-effector allows accurate positioning, adjusting, and fixation of the bricks. (Image: Construction Robotics.)

than the original SAM prototype. SAM also includes an operation and path planning software (Figure 2.105) and an on-site control panel interface (Figure 2.106).

For further information, see Construction Robotics (2015).

2.8 Concrete Distribution Robots

Concrete distribution robots are used to distribute mixed concrete with uniform quality over large surfaces or over formwork systems and thus partly or fully automate a highly repetitive and labour-intensive task. The speed robots can add to

Figure 2.104. Side view of SAM in operation. (Image: Construction Robotics.)

Figure 2.105. SAM operation and path planning software. (Image: Construction Robotics.)

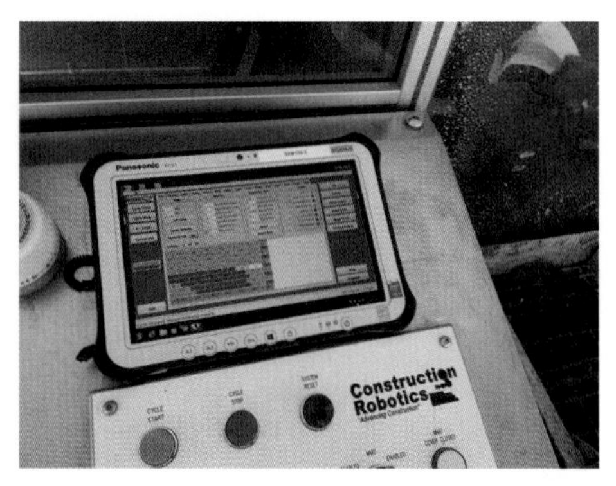

Figure 2.106. On-site control panel for SAM. (Image: Construction Robotics.)

concrete distribution is also an important factor. The supply of concrete to a work area can be efficiently channeled using concrete pumps. However, the distribution of concrete by human workers can become a bottleneck operation slowing down the overall flow of work and material. In contrast, the use of high-performance robots is complementary with the use of high-performance concrete supply pumps. Systems in this category range from horizontal and vertical logistics/supply systems to compact mobile concrete distribution and pouring systems that operate on individual floors as well as larger, stationary systems. In particular, the concrete distribution and pouring systems conduct simple and predefined (e.g., swaying) motions in a repetitive and accurate manner that is able to distribute the concrete uniformly. Most systems automate these simple repetitive motions, and overall guidance through the work area through human intervention is needed, thus indicating hybrid assistive systems rather than fully automated solutions.

Concrete Distribution Robots		
43	Stationary concrete distribution robot	Obayashi and Mitubishi
44	DB Robo concrete distributor	Takenaka
45	Horizontal concrete distribution robot	Takenaka
46	Automated on-site concrete logistics system	Konoike,
47	Mobile concrete distribution robot	Tokyu Construction Co., Ltd.
48	Concrete distribution robot	Kajima
49	Rail-guided ground-level concrete distributor	Kajima

System 43: Stationary Concrete Distribution Robot

Developer: Obayashi and Mitsubishi

This stationary concrete distribution robot (Figures 2.107 and 2.108) was jointly developed by the contractor Obayashi and the construction equipment producer

Figure 2.107. Side view of stationary concrete distribution robot. (Image: Obayashi Corporation.)

Figure 2.108. Front view of stationary concrete distribution robot. (Image: Obayashi Corporation.)

Mitsubishi. The robot can simplify the distribution logistics and pouring of the concrete on the construction site. Furthermore, the robot can be used for lifting material or equipment into place. The robot has a relatively large working space compared to other concrete distribution robots. However, its towering crane-like structure, along with the fact that it is not integrated with a mobile platform, make it relatively inflexible compared to other robots and unsuitable inside of a building. The robot system consists of three sections: a base part that provides stability to the robot and fixes it to the floor on which it operates, a middle section composed of four joints and links spanning the working space of the robot, and an end-effector section. Within the end-effector section, an end-effector for pouring concrete and an end-effector for lifting material are installed. Furthermore, a concrete pump and a tubing system for concrete "logistics" are included in the system. Further specifications of the system are given in Table 2.3.

Table 2.3. *Stationary Concrete Distribution Robot System Specifications*

Item	Specifications
Boom	
Type	Automatic control, full hydraulic multijoint with distributor
Max. length and height	32 m(L) and 53 m (H)
Rotation angle	400° (for boom) and 140° (for distributor)
Concrete pipe diameter	0.125 m
Control method	Automatic as well as manual control
Power unit	27 Mpa × 60 l/minute 37 KW
Control unit	Micro-computer and hydra electric servo valve

Source: Council for Construction Robot Research (1999, pp. 38–39).

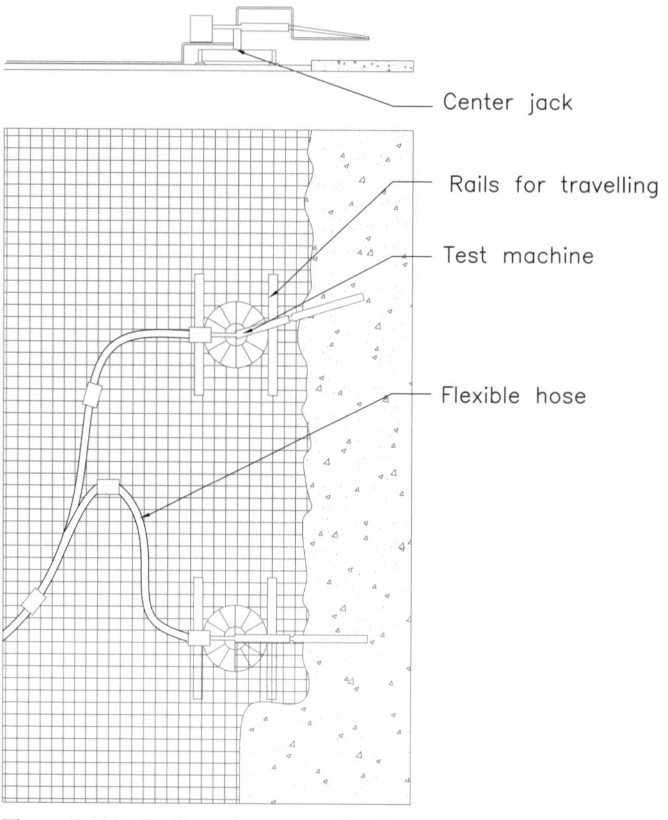

Figure 2.109. Outline of concrete distributor system components. (According to Construction Robot System Catalog in Japan, 1999, pp. 80–81.)

System 44: DB Robo Concrete Distributor

Developer: Takenaka Corporation

Concrete distribution on individual building floors is usually carried out by human workers because of the inaccessibility of such areas using larger and powerful distribution equipment. The DB Robo concrete distributor is a small (dimensions: W 5.05 m × L 3.60 m × H 1.69 m) and relatively lightweight robot that can move on individual floors and distribute the concrete automatically in a preprogrammed manner (Figure 2.109). The robot consists of a mobile base, a hose through which concrete is pumped, a manipulator that brings the hose (which is guided over the manipulator) into a rough position, and an end-effector that controls the final position of the hose. The manipulator makes it possible to sway the hose and thus to distribute the concrete in a certain pattern around the robot. Because of the compactness of the systems several robots can be used on one floor in parallel to speed up the concrete distribution work. The DB Robo concrete distributor system can be remotely supervised by one skilled worker.

For further information, see Council for Construction Robot Research (1999, pp. 80–81).

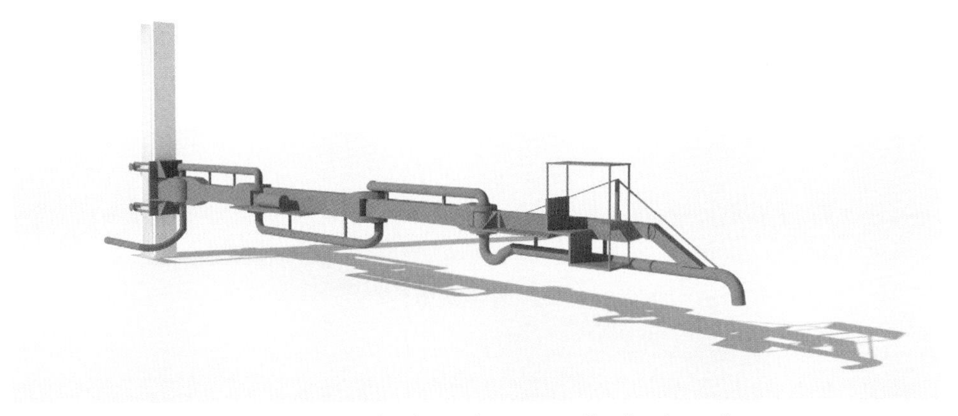

Figure 2.110. Outline of Takenaka's horizontal concrete distribution robot.

System 45: Horizontal Concrete Distribution Robot

Developer: Takenaka Corporation

This robot was introduced to improve and automate concrete distribution on individual floors. The robot has a relatively large work space and can cover a work radius of more than 20 metres around the location where it is positioned (Figures 2.110 and 2.111). It can be fixed either to a column or moved on the site by a mobile platform. The main body of the robot consists of a manipulator that can be lifted up and down (1 DOF) vertically and that consists of four segments connected horizontally through joints (thus providing another 4 DOF) as well as an end-effector

Figure 2.111. Takenaka's horizontal concrete distribution robot in action. (Image: Takenaka Corporation.)

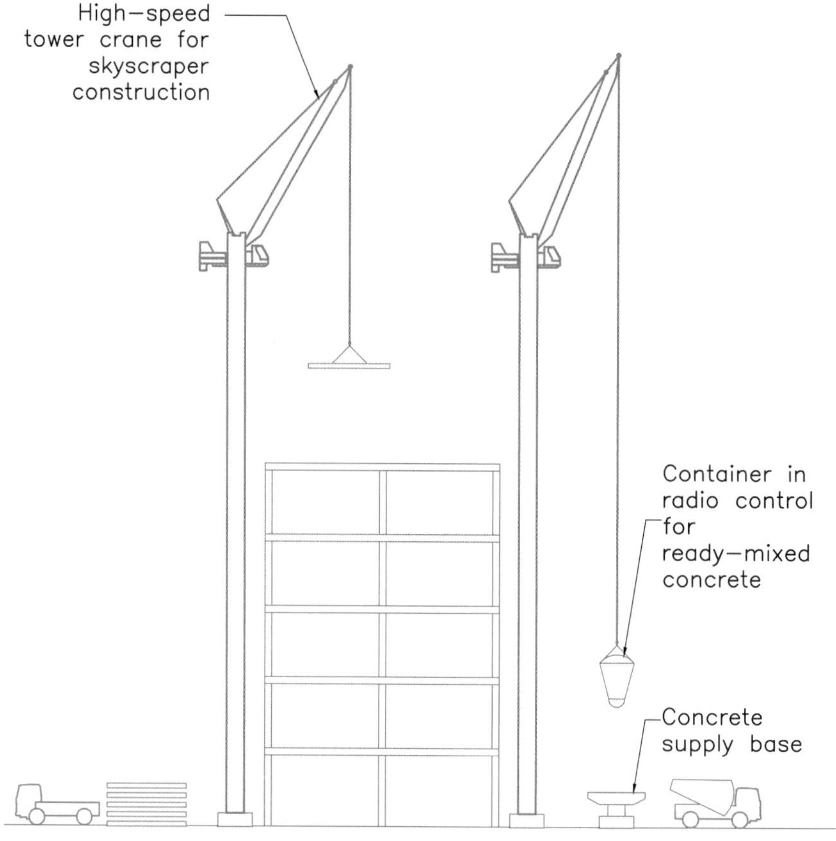

Figure 2.112. Overview of Konoike's automated concrete distribution system. (Adapted from Council for Construction Robot Research (1999, pp. 180–181).

that can fine position and sway the hose through which the concrete is pumped in a certain pattern. Takenaka showed that the system is able to reduce the required human labour by about 30%. The robot can be operated in an automatic mode and in a manual mode. The manual mode is used in areas containing a series of obstacles, making the correct positioning of the end-effector difficult. The automatic mode is used in areas with relatively view and large obstacles.

For further information, see Council for Construction Robot Research (1999, pp. 78–79), Sherman (1988), and Aoyagi and Shibata (1988).

System 46: Automated On-Site Concrete Logistics System

Developer: Konoike Construction Co. Ltd.

Konoike's automated concrete distribution system (Figure 2.112) is a solution for the optimization of concrete logistics on construction sites. It is particularly useful for very large or high-rise construction sites where concrete pumps are difficult to use. The system consists of two robotic cranes (modified standard cranes that were equipped with encoders), the climbing mechanism of the cranes, a concrete lifting container with integrated sensors and communication systems, and a concrete supply base. The concrete supply base can be considered a type of fixture, which standardizes

Figure 2.113. Mobile concrete distribution robot. (Image: Tokyu Construction Co., Ltd.)

the locations for feeding and picking up the concrete-lifting containers on the ground level. This structures the supply of concrete from the concrete lorry to the lifting container. Concrete supply to the construction site as well as to the robot system is carried out by ready-mixed concrete lorries just in time and just in sequence. Therefore, the software system developed along with the robotic crane system allows integration with the supply chain, a real-time on-site monitoring system, and a programming and control interface.

For further information, see Construction Robot Research (1999, pp. 180–181).

System 47: Mobile Concrete Distribution Robot

Developer: Tokyu Construction Co., Ltd.

The mobile concrete distribution robot consists of two four-wheeled mobile, robotic vehicles over which a hose (through which concrete is pumped) is synchronously guided and positioned (Figure 2.113). In a cooperative process, one of the vehicles is used only to guide the hose in place, and the second one contains the actual concrete pouring end-effector, which can be moved up and down as well as left and right by a winch mechanism. Through the usage of these two robotic vehicles working synchronously, concrete pouring on the floors of a large building with large floor areas can be assisted and to a certain extent automated. The wheels are equipped with treaded rubber tires to be able to move on the already installed reinforcing bars or meshes. Once the concrete is dry and ready, additional finishing is required. The robot can be operated in automatic mode as well as directly controlled and physically guided by construction workers.

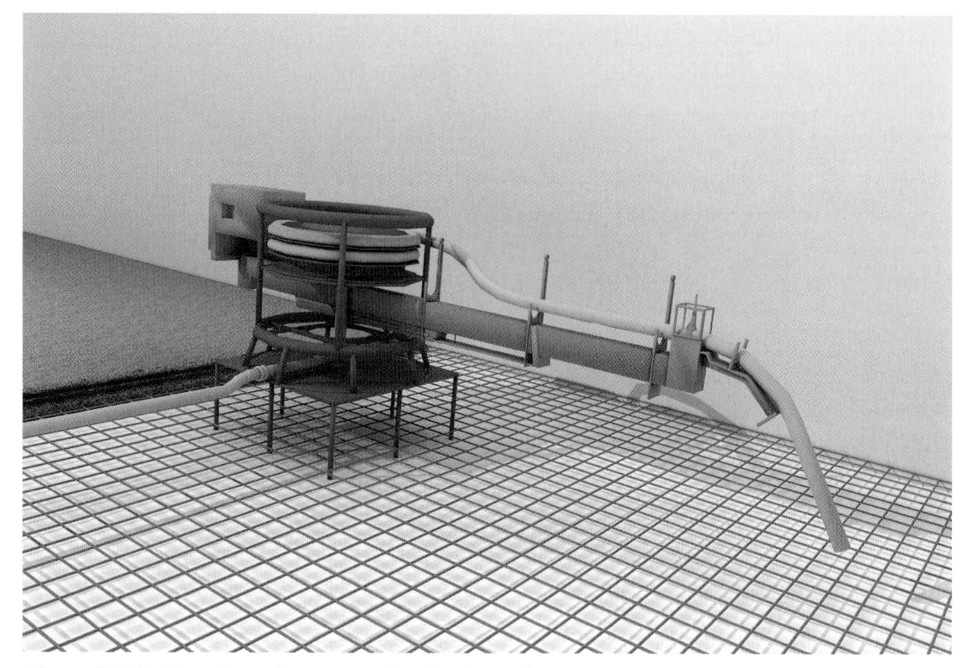

Figure 2.114. Overview of concrete distribution robot system.

System 48: Concrete Distribution Robot

Developer: Kajima Corporation

Kajima developed this robot for automating the distribution of concrete within building structures (Figures 2.114 and 2.115). The robot system is compact and can be repositioned by cranes, mini cranes, or other light transport vehicles on the construction site. The robot system consists of a repositionable stationary platform, a rotation mechanism installed on this platform, a boom, an end-effector, a hose system, and a concrete pumping system for steady concrete supply. The end-effector can adjust the position of the hose end that releases the concrete. The robot system can either be preprogrammed to distribute the concrete in a given pattern or it can be guided manually in conjunction with a construction worker.

System 49: Rail-Guided Ground-Level Concrete Distributor

Developer: Kajima Corporation

On large construction sites, enormous amounts of concrete and other material need to be moved. For this reason, Kajima developed a rail-guided ground-level concrete distribution system that can be installed along main distribution lanes (Figure 2.116). The system consists of a rail system, a mobile platform, and a material loading and unloading system installed on the mobile platform. The system's efficiency can be maximized by using on-site logistics routes that are concentrated and standardized. In the event that the system is connected to a supply of material on its input end and to a fully integrated and/or automated processing system on its output end, a production line–like material flow can be established on the construction site. The guiding rails make the system comparatively complex to set up, but on the other hand, once set up, the system guarantees high productivity.

Figure 2.115. Concrete distribution robot system in action. (Image: Kajima Corporation.)

Figure 2.116. Rail-guided ground-level concrete distributor. (Image: Kajima Corporation.)

2.9 Concrete Levelling and Compaction Robots

Similarly to practices during the 1980s, currently, a considerable amount of concrete structures are cast on-site. Automated or robotic concrete levelling and compaction systems are able to ensure uniform compaction (and thus concrete quality), and if properly integrated with other operations, allow enhancement of the overall speed of concerting works. Concrete levelling and compaction are frequently demanded and closely linked work activities on the construction site. Concrete levelling is the process of levelling the poured or roughly distributed concrete to have a more compacted and planar (but not finished) concrete layer. It involves a large number of repetitive operations (e.g., skimming with slide dampers). If manually done, continuous input of strong physical power is necessary, which limits the operational speed, making conventional labour-based concrete levelling a time- and resource-intensive construction operation. Automation of levelling operations – similar to concrete finishing operations – speeds up the levelling process, enhances labour productivity, and maintains a highly uniform quality for the entire surface. Concrete compaction removes air from the concrete and compacts the particles inside the concrete mixes, which enhances the density and strengthens the bonding weaknesses between the concrete and reinforcement. Systems in this category in most cases both level and compact the concrete. However, whereas the systems of Mai, Fujita, Takenka, and Lomar focus more on the screeding/levelling process (which also compacts the concrete to a certain extent), Takenaka's concrete floor compaction system is specialized in compacting the concrete but at the same time also levels the concrete to a certain extent. Obayashi's concrete wall compaction system makes it possible to automate, simplify, and speed up the compaction process for formwork-based wall concreting.

Concrete Levelling and Compaction Robots		
50	Concrete floor compaction system	Takenaka
51	Concrete wall compaction system	Obayashi
52	Automatic compacting and levelling system	Mai International
53	Floor levelling robot CALM	Fujita
54	Automatic floor screeding robot	Takenaka
55	Screeding robot LOM 110	Lomar SRL

System 50: Concrete Floor Compaction System

Developer: Takenaka Corporation

Concrete floor compaction is a critical process that needs to be done within a limited time following pouring. Therefore, speed and power are two important factors when thinking about automating or robotizing this process. The concrete floor compaction system, developed by Takenaka (Figure 2.117), consists of a main body through which two plates are activated, creating a walking motion allowing the robot to move over an area of previously poured concrete. In addition, the two plates produce slight vibrations and thus apply a dedicated pressure to the surface. The robot thus is able to press out water from the concrete, compact it, and promote the hardening process. The robot is efficient in terms of the number of people required,

Figure 2.117. Concrete floor compaction system.

the quality, and the time needed to complete the task. Furthermore, the efficiency of the system increases with the size of the floor area to be treated.

System 51: Concrete Wall Compaction System

Developer: Obayashi Corporation

An automatic concrete vibrator was developed by Obayashi to improve the quality of work, reduce manpower, and create a better working environment. Concrete needs to be compacted right after it is poured into the formwork of building elements such as columns, beams, slabs, walls, and so forth on the construction site to avoid quality problems in the concrete and a decrease of the compression strength of the concrete. Furthermore, in concrete formwork, there are spaces where steel reinforcing bars are too close to each other and conventional mechanical rod type vibrators have only a limited impact. The system consists of a set of vibrators that can be attached to the outside of the formwork, as well as a central, mobile power supply and control module (Figures 2.118 and 2.119). When the concrete is poured into the formwork, sensors (concrete level sensors, sensitive to alkali in the concrete, in the shapes of galvanized 25- to 30-mm steel nails integrated in the formwork) the system starts automatically to work. Applying a dedicated, modulated vibration to the formwork during pouring and for a certain time after pouring promotes even distribution, settlement, and hardening of the concrete in the formwork. Once the process is completed, the attached vibrators can be removed from the surface of the formwork and brought to the next location. The whole concrete compaction system is supervised remotely from a portable PC.

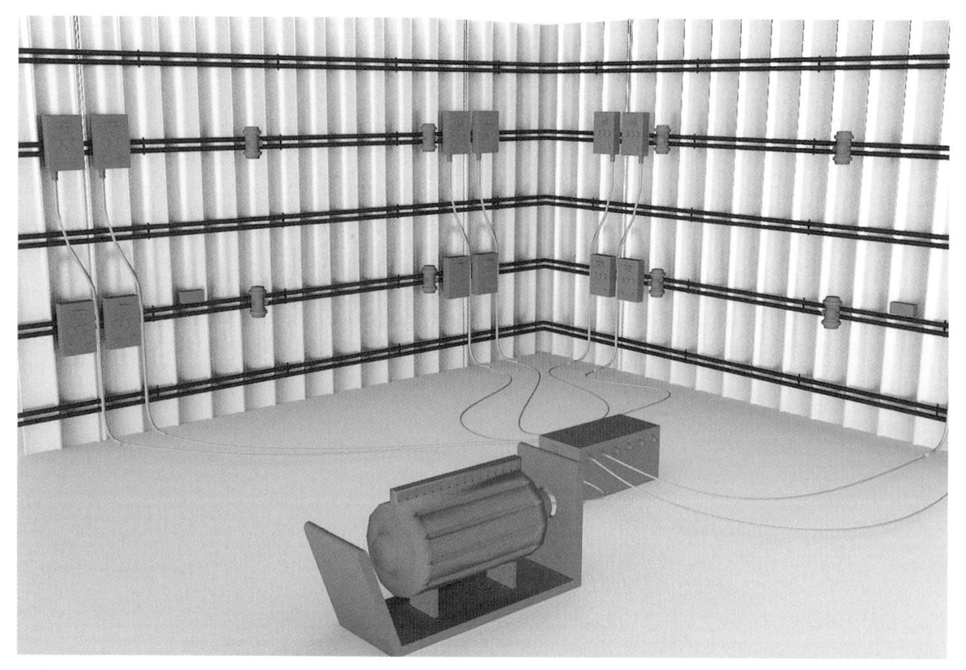

Figure 2.118. Overview of concrete wall compaction system.

Figure 2.119. Concrete wall compaction system in action. (Image: Obayashi Corporation.)

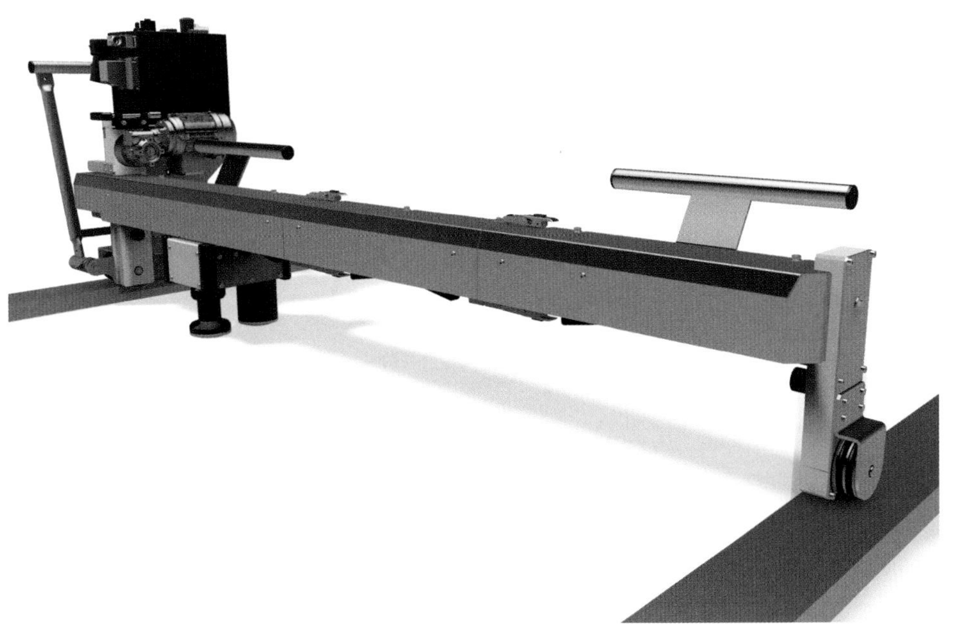

Figure 2.120. The gantry-type main body of the system carrying the end-effectors moves on rails along the work area. (Image: MAI® International GmbH.)

For further information, see Council for Construction Robot Research (1999, pp. 76–77).

System 52: Automatic Compacting and Levelling System: MAI® 2Floor Master

Developer: MAI® International GmbH

Screeding is a physically demanding task for construction workers, particularly when large concrete areas have to be processed. MAI's automatic compacting and levelling system performs both compacting and levelling partly automatically. It still requires that the site workers install it, supervise its operation, and prepare the concrete in the processing area; however, the core task is automated, reducing the need for human labour and ergonomically unsuitable work postures and enhancing quality and precision at the same time. The width of this system is adjustable up to a width of 4 metres. The gantry-type main body of the system (Figures 2.120 and 2.121) carrying the end-effectors moves on rails that have to be installed (and that thus have to be considered in the construction process) along the area that is intended to be processed. The compaction and levelling end-effector of the system is driven by the gantry-type main body of the system and moves across the width of the workspace and orthogonal to the rails. During the movement, one part of the end-effector (front side facing the unprocessed area) compacts the concrete and at the same time the second part of the end-effector (backside facing the processed area) smoothens the concrete. The system can be supervised by one worker (Figure 2.122) and processed under optimal conditions up to 110 m^2/hour. It consists of lightweight modular components that make assembly, disassembly, and transport of the system easy.

Figure 2.121. Depiction of the mechanical approach. (Image: MAI® International GmbH.)

Figure 2.122. The system can be operated on the site by one worker. (Image: MAI® International GmbH.)

Figure 2.123. Overview of concrete floor levelling robot system.

System 53: Concrete Floor Levelling Robot (CALM)

Developer: Fujita Corporation

The concrete floor levelling robot (CALM; Figures 2.123 and 2.124) by Fujita was developed with the aim of reducing the number of skilled workers during the concrete levelling process, increasing construction speed, and enhancing the quality

Figure 2.124. Concrete floor levelling system in action. (Image: Fujita Corporation.)

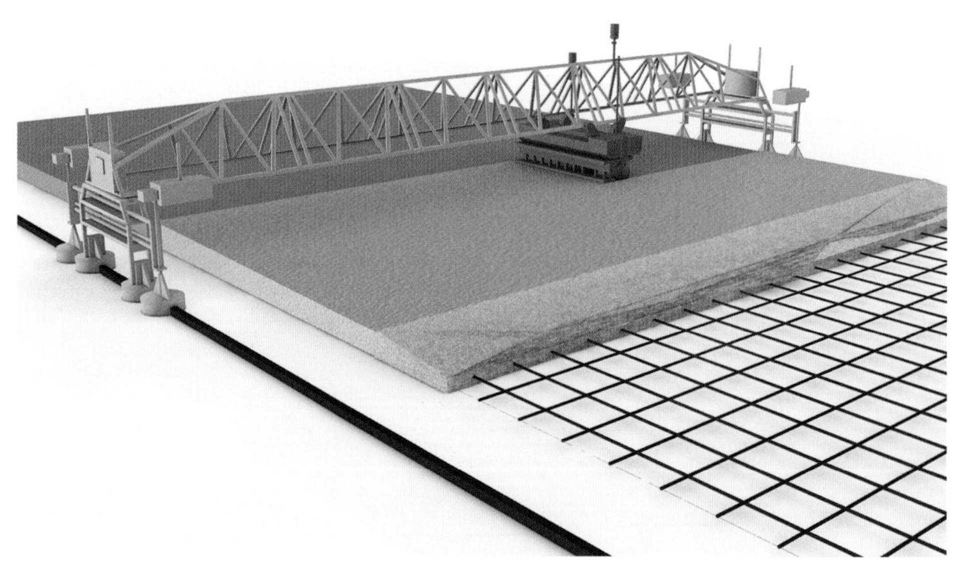

Figure 2.125. Overview of automatic floor screeding robot system.

and accuracy of the surfaces created. This is a monotonous and repetitive task. Also, the quality demand of this work is high. The system consists of preinstalled, temporary guiding rails (which act both as the formwork for the concrete and the guiding element for the machine), the robot's main body overspinning the area between the rails moving on top of the rails, and a levelling end-effector at the bottom of the main body. Several sensor systems (including conventional cameras, spectrum cameras for quality control, laser equipment, and an automated positioning system) are mounted onto the frame of the main body. The main body carrying and applying the end-effector is activated by electric motors and moves horizontally on the rails to level the concrete surface. It can also be moved vertically up and down to adjust height level. The speed of movement and the required level can be adjusted to desired conditions.

System 54: Automatic Floor Screeding Robot

Developer: Takenaka Corporation

Concrete screeding follows the process of concrete pouring. Its main purpose is to remove the excess concrete, level the concrete, smooth the concrete surface across the working area, and to a certain extent promote the compaction and hardening of the concrete. The system follows a gantry-type structural approach (Figure 2.125) and consists of guiding rails, a height-adjustable frame that spans the work area and that can move along the guiding rails, an end-effector that is positioned in a rectangular orientation with respect to the frame and that moves along the bottom of the frame over the work area, and a set of sensors for navigation and work process and quality control. Once the end-effector width (1.6 m) has moved over the work area and completed a segment, the frame is guided by the rails, repositioned automatically, and the process starts again. The screening end-effector uses a 1.6-metre rotating cylinder for removing excess concrete and a vibrating plate for smoothing the concrete surface. To achieve optimum results, the robot uses a laser scanning system as the main sensor

Figure 2.126. Screeding robot LOM 110. (Image: LOMAR.)

system for levelling height adjustment. The frame can be extended extend up to 17 metres, thus making it possible to span and screed an area with a width of up to 15.5 metres.

For further information, see Council for Construction Robot Research (1999, pp. 250–251).

System 55: Screeding Robot LOM 110

Developer: Lomar SRL

The LOM 110 is the newest product of a series of automatic floor screeding tools developed by the Italian company Lomar. The robot consists of a mobile, tracked base platform with an attachable/detachable manipulator (Figure 2.126). This manipulator contains the screeding end-effector as well as a part of a laser-guided system that is the basis for adjusting height and movement of the manipulator (Figure 2.127). The manipulator performs a simple but robust and accurate lateral movement. The tracked mobile platform is responsible for moving the manipulator and end-effector forward. Part of the robot system is a separate stilt with a laser guidance system that can be placed in the room in which the robot is supposed to work and that provides an external reference point for navigation and height adjustment of the manipulator.

Figure 2.127. Part of the robot system is a separate stilt with a laser guidance system. (Image: LOMAR.)

The robot is compact enough to fit through doors and other small openings and it can be disassembled for conveyance to the operational floor.

2.10 Concrete Finishing Robots

Floor finishing, as seen from an ergonomic viewpoint, is one of the most critical construction processes. Construction workers carry or guide trowel smoothers for several hours in a bent posture. To relieve construction workers from this work while maintaining consistent construction quality, various companies have developed and deployed concepts for robots able to execute that task. A few such robots include Shimizu's Flat-KN (400–800 m^2/hour), Kajima's KoteKing (about 500 m^2/hour), Takenaka's Surf Robo (up to 300 m^2/hour), Obayashi's mobile floor finishing robot (about 500 m^2/hour), Hazama Ando's mobile floor trowelling robot (approximately 300 m^2/hour), and Kajima's mobile floor finishing robot (more than 300 m^2/hour).

In comparison, Cousineau and Miura (1998) stress that in Japan, a skilled labourer is not considered to be able to complete more than 120 m^2/hour by conventional methods. After the intensive research and development phase ended in 1985, the first concrete finishing/smoothing robots were commercially deployed in 1986 to assist in the finishing of concrete floors of larger buildings, high-rises, power plants, and other commercial buildings. The innovation and commercialization process has continued since then. For example, the Kote King system was finally outsourced in 1995 to Tokimec, which then developed the Robocon system version. The Robocon system is still in use today and is lighter (Kote King weighed 141 kg; Robocon, 68 kg), quieter (Kote King, 70 db; Robocon, 50 db), more productive (Kote King, 500 m^2/hour; Robocon, 800 m^2/hour), and less costly (Kote King, €40,000; Robocon, €20,000).

Each of these STCRs is able to operate in a predefined pattern and one that is suitable for the robot's mobility system on an individual floor. The unit providing the mobility is either docked as a separate module to the finishing or smoothing unit (as is, e.g., the case with Obayashi's robot) or directly integrated into the finishing or smoothing mechanism. Most systems utilized end-effectors that could be equipped with different types of rotation tools such as rotating trowel blades or discs that are rotated and pushed with a certain controllable pressure towards the surface. A few systems worked with alternative approaches utilizing controllable vibrating blades or elements (e.g., Hazama). Energy is either supplied through an on-board generator (making the system heavier but easier to deploy) or by an external power source through ceiling-suspended cable systems (making the systems lighter and more dexterous but requiring more complex temporary installations).

The operation modes range from direct remote control to automated navigation and obstacle avoidance along preprogrammed general routes. In many cases, gyroscopes and laser scanners assisted navigation and motion planning on a low level within the preprogrammed travelling routes. The setup of STCRs on a floor or site, or the transfer of the robot from one floor to another, is a general weakness of STCRs. The Hazama system provided a solution to that problem: the robot is not only the lightest in this category, but also allows for dismantling into five 20-kilogram modules that can easily be transported by human workers on the site.

Concrete Finishing Robots		
56	Robotic concrete floor trowel	Robotus
57	Floor trowelling robot Robocon	Tokimec
58	Floor trowelling robot	Hazama
59	Concrete-floor finishing robot	Kajima
60	Concrete (interior) floor finishing robot Flat-Kun	Shimizu
61	Concrete-floor finishing robot Surf Robo	Takenaka
62	Automatic laser beam–guided floor finishing robot	Obayashi
63	Concrete (exterior) floor finishing robot Flat-Kun	Shimizu
64	Concrete-floor smoothing robot	Takenaka
65	Mobile floor finishing robot	Kajima

System 56: Mobile Robotic Concrete Floor Trowelling System

Developer: E. Tambakevicius, S. Atkociunas, M. Malcius, M. Muncys, T. Brazauskas, K. Valentoniene, M. Tambakevicius

This robot was developed to execute the concrete finishing task with high quality and minimum human supervision, as well as to relieve construction workers from this repetitive work (Figure 2.128). The robot principally adds automatic, robotic control to a conventional hydraulic power trowel machine. The system is driven by a 45 HP turbo diesel engine and equipped with a navigation system based on laser scanning that allows the robot to recognize the work area, including obstacles both in indoor and outdoor environments. The laser scanner is also used to measure levels, inclinations, and inaccuracies of the floor to be processed. The operator can adjust the concrete area to be finished, working trajectories, and other parameters such as trowel speed, as well as monitor the progress of the work through a remote control interface (connected to the robot by LAN or WLAN), and the robot can then perform the task autonomously.

Figure 2.128. Robotic concrete floor trowel. (Image: Evaldas Tambakevicius.)

The outlined information was provided by Mr. Evaldas Tambakevicius. For further information, see Robotus (2015).

System 57: Mobile Floor Trowelling Robot Robocon

Developer: Tokimec

The robotic floor-trowelling robot developed by Tokimec, also known as Robocon, is another robotic version of the manually controlled concrete trowelling system (Figure 2.129). It is relatively small in size compared, for example, to the concrete floor trowel robot developed by Robotus. This affords an advantage while working on individual floor levels inside buildings and makes it possible for the robot to perform its tasks efficiently, regardless of how complex the work area is. The robot has a built-in navigation system, which enables it to work fully autonomously while avoiding obstacles. The robot can also be controlled remotely by one supervisor to perform certain tasks as needed. The robot uses two trowelling end-effectors working in parallel to perform the tasks. Thanks to its relatively small size and light weight that is evenly distributed on a wide working area underneath, the robot can perform the floor trowelling on concrete floors in early hardening stages.

For further information, see Taylor et al. (2003).

System 58: Mobile Floor Trowelling Robot

Developer: Hazama Ando Corporation

Hazama's floor trowelling robot is able to automate the finishing of large concrete floors. The size of the robot system (L 2.0 m × W 0.8 m× H 0.9 m) and its low weight allow it to be operated on individual floors within constructed structures of all types. The robot system consists of a mobile platform centre unit with an electric motor on board, working units flanking the centre unit, and a measuring and

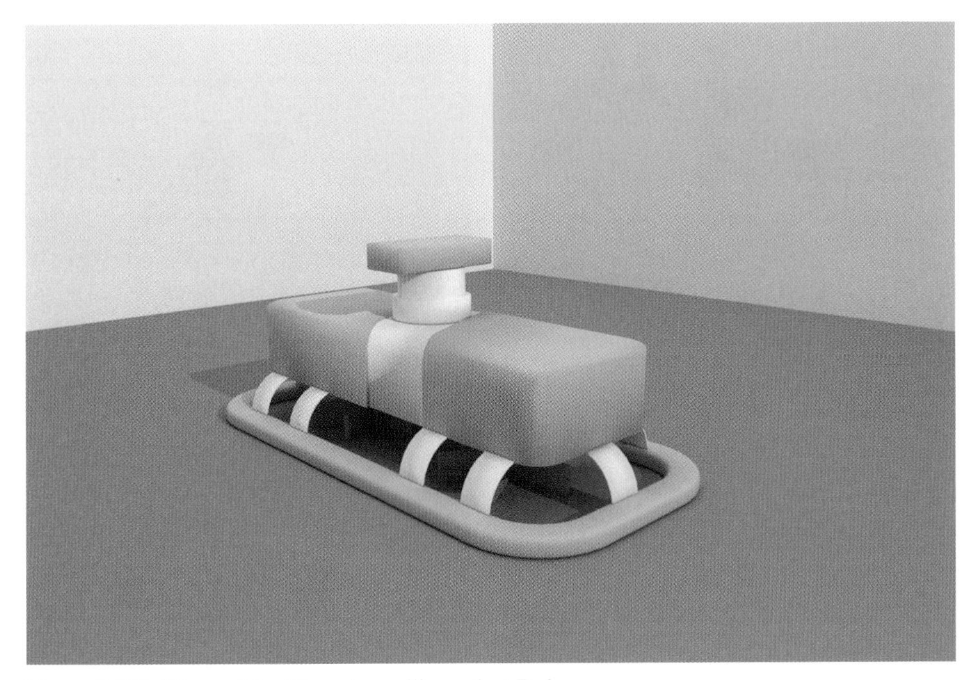

Figure 2.129. 3D model of floor trowelling robot Robocon.

control unit placed on top of the mobile platform centre unit (Figures 2.130–2.133). Mounted on the bottom of the working units are the end-effectors conducting the floor finishing work. The end-effectors utilize ultrasonic vibration units linked to the trowels as opposed to the rotating trowels that most other floor finishing robots use.

Figure 2.130. Mobile floor trowelling robot system in action, side view.

Figure 2.131. Mobile floor trowelling robot, front view.

Figure 2.132. Main mobile floor trowelling robot unit with mobility systems and sensors.

Figure 2.133. End-effector for mobile floor trowelling robot.

Figure 2.134. Mobile concrete floor finishing robot system in action, front view. (Image: Kajima Corporation.)

The robot can be disassembled into five modules, which can easily be carried by a human worker to the floor or location where the robot shall operate. The moving path and finishing sequence of the robot can be preprogrammed. Alternatively, the robot can be either fully or partly remote controlled (e.g., emergency stops, etc.).

For further information, see Council for Construction Robot Research (1999, pp. 224–225).

System 59: Mobile Concrete Floor Finishing Robot

Developer: Kajima Corporation

The concrete floor finishing robot from Kajima (Figures 2.134 and 2.135) was built to perform the final labour-intensive stages of finishing and smoothing a concrete slab. The robot consists of a mobile platform (drive module mounted atop three rollers) and an end-effector (rotary finishing trowel that is pulled behind the mobile platform). The end-effector consists of activated and adjustable rotating concrete finishing trowels that produce the desired finishing. The driving module is equipped with a gyroscope and a linear distance sensor that allow the robot to navigate autonomously within a predefined work area and complete work on a floor (except for a stripe of 30 cm around walls and other obstacles). Work initiation and termination

Figure 2.135. Detailed view of trowelling end-effector at the back side.

(including emergency stops) can be controlled remotely. The light weight of the robot allows ease of transportation and implementation on floors that may have lesser load capacities. Safety features of the robot include a touch sensor system (a 30-cm bump-sensitive frame around the front of the robot) for obstacle detection and an alarm noise emitted during operation. As the robot has no on-board battery pack, an external power supply through a cable system that is flexibly suspended from the ceiling along the work trajectory is required.

Further specifications of the robot:

- Dimensions: L 1.150 m × W 1.225 m × H 0.646 m
- Weight: 141 kg
- Max travel speed: 18 m/minute
- Max working capacity: 500 m^2/hour
- Power supply: 200 VAC, three phases, 1.5 kVA

For further information, see Council for Construction Robot Research (1999, pp. 230–231).

Figure 2.136. Flat-KN in action on the building floor. (Image: Shimizu Corporation.)

System 60: Mobile Concrete Floor Finishing Robot Flat-KN

Developer: Shimizu Corporation

Conventionally, concrete pouring, screeding, smoothing, and finishing require a huge amount of manual handling, and the work can be physically strenuous. The Flat–KN (Figures 2.136 and 2.137) consists of three sets of trowel end-effectors located around a mobile robotic platform, a controller system that allows the robot to be either manually controlled or remotely operated, and an anticollision device. The trowels are attached on the end of each retractable arm structure. The blade pressure against the floor can be adjusted in response to the hardness of the surface to be finished. The system is equipped with a portable control interface that is wirelessly connected to the robot allowing the operator to control remotely operation speed and orientation of the robot. In case of emergency, the touch sensor will guide the machine to avoid obstacles and switch off the machine automatically when necessary.

According to Shimizu, the concrete flooring finishing process can be divided into four stages that can be conducted by the robot system using different types of end-effectors: (1) primary finishing, (2) first finishing, (3) second finishing, and (4) final finishing. During the primary finishing stage, the trowel blades are disassembled and exchanged with levelling disks used to level and flatten the floor surface. The smoothing task is carried out during the first finishing stage by switching the levelling disks back to trowel blades. The blades are rotated as slowly as possible to protect the concrete surface. During the second finishing stage, the blades are adjusted to a faster speed to conclude the smoothing task. During the final stage, the robot inspects the

Figure 2.137. Flat-KN variation used in an outdoor environment. (Image: Shimizu Corporation.)

surface and thus helps to assess the quality of the work done. A disadvantage of the Flat-KN is that it is difficult for the system to properly finish corners and edges between walls and floors because of its rounded shape, and these areas must be completed manually.

Further specifications of the robot:

- Trowel blade size: L = 0.29 m, W = 0.15 m
- Trowel blade angle: 0–10°
- Outer diameter = 2.30 m, H = 0.81 m
- Travel rate = 0–10 m/minute
- Weight = 300 kg
- Trowel rotating speed: 70–100 rpm
- Trowel revolving speed: 0–13 rpm
- Productivity: 400–800 m^2/hour, four to eight times faster than manual speed.

For further information, see Council for Construction Robot Research (1999, pp. 246–247).

System 61: Mobile Concrete Floor Finishing Robot Surf Robo

Developer: Takenaka Corporation

The Surf Robo, developed by Takenaka, provides a compact, flexible, safe, and efficient robot for concrete floor finishing (Figures 2.138 and 2.139). The robot consists of a mobile platform, two trowelling end-effectors, a remote control operation

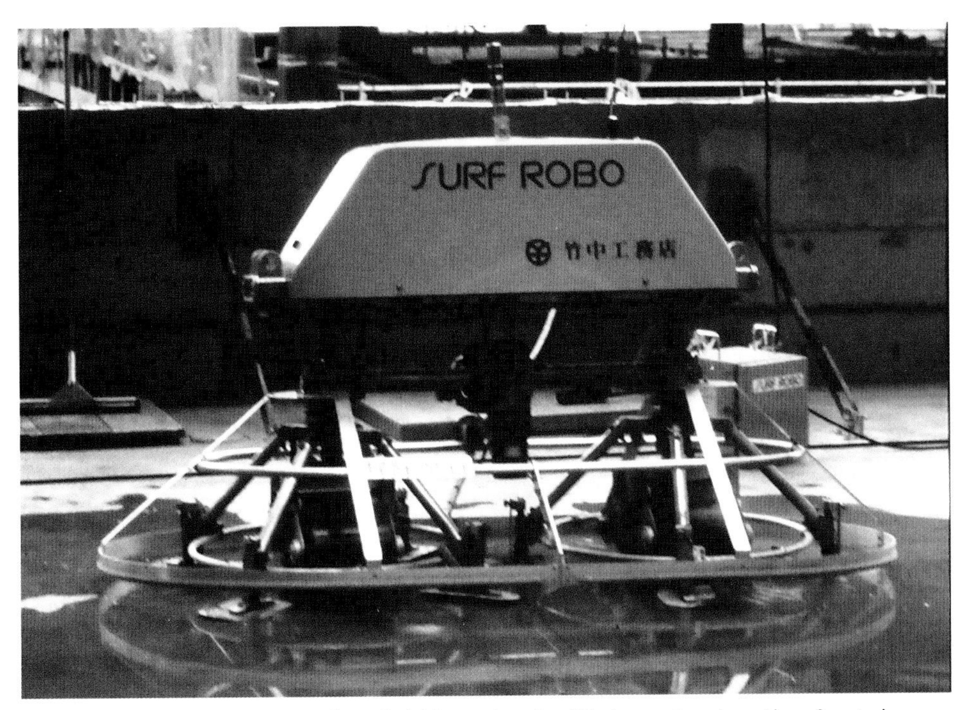

Figure 2.138. Mobile concrete floor finishing robot Surf Robo system in action, front view.

panel, a power supply unit that provides electricity to the robot through a flexible ceiling suspended system, and touch-sensitive safety frame allocated around the main body and the end-effectors. Locomotion is provided through two caterpillar-like end-effectors each allocated in the centre area of the trowel end-effectors. Each of the two trowel end-effectors contains four trowels. The speed of rotation as well as the pressure applied to the concrete surface by those trowels can be adjusted according to the applicable concrete curing stages. The blades can be disassembled and cleaned separately. The robot moves forward with its length oriented perpendicular to motion trajectory. Work area stripes thus processed by the robot need to overlap slightly to guarantee a complete processing of the area. The robot can be both remotely controlled and operated in automatic mode. Under the automatic control mode, the travel path is preprogramed following a previously software determined optimized travel sequence. The safety frame is equipped with touch sensors to prevent collisions during system operation and a warning light system is installed on top of the main body, which will indicate if an error has occurred as well as the status of the work carried out by the robot. To reduce the weight of the system and improve manoeuvrability during operation, an overhead cable system links it with the power supply unit. The overhead cable system is a subsystem that needs to be installed by human workers prior to the operation of the robot.

Further specifications of the robot:

- Dimensions: L = 2.23 m, W = 1.25 m, H = 1.35 m
- Weight: 185 kgf
- Contact pressure when trowelling: 0.12 kgf/cm^2
- Finishing work capability: 300 m^2/hour (with two-step finishing)

Figure 2.139. Mobile concrete floor finishing robot Surf Robo system in action, top view.

- Finishing width: about 2.14 m
- Rotation speed of blades: 0–35 rpm
- Speed 0–12 m/minute

For further information, see Council for Construction Robot Research (1999, pp. 252–253).

System 62: Mobile Floor Finishing Robot

Developer: Obayashi Corporation

The robot can be divided into a mobile unit and a detachable end-effector unit (Figure 2.140). The mobile unit contains a battery pack that provides power for the activators and sensors. When the battery pack needs to be recharged the end-effector unit can be detached and attached to another mobile unit, thus guaranteeing uninterrupted operation. In contrast to other systems, the mobile floor finishing robot does not require a power supply through a complicated cable suspension system. Obayashi also intends to develop further end-effector units (e.g., for inspection, cleaning, etc.)

Figure 2.140. Mobile floor finishing robot system in action. (Image: Obayashi Corporation.)

so that the mobile unit can then be equipped with task-specific modules. The robot can be operated both manually using a personal computer and automatically via preprogramed instructions. A sophisticated navigation system allows the robot to prevent errors or delays during automatic operation. The navigation system consists of three subsystems: (1) a laser-based navigation system, (2) a self-position detector (embedded into the robot), and (3) "corner cubes" (installed around the work area). The laser navigator and the self-position detector detect light beams reflected back from the corner cubes to determine the robot's position in the work area. Based on sensed obstacles, the work area, determined by the corner cubes and other input data provided, the control software automatically generates an optimized working path. The robot's trowelling system can be calibrated towards various on-site conditions and necessities (e.g., type of concrete, concrete curing conditions, demanded robot operation time, etc.). Therefore trowelling speed and exact trowelling end-effector position can be adjusted accordingly.

Further specifications of the robot:

- Dimensions: L = 1.985 m, W = 1.560 m, H = 1.100 m
- Control mode: Manual and automatic
- Sensor: Opening and obstacle detective sensor and laser positioning sensor
- The finishing capacity is 500 m^2/hour at a running speed of 0–11 m/minute, with continuous operating time of 4 hours or longer.

Figure 2.141. Shimizu's concrete exterior floor finishing robot. (Image: Shimizu Corporation.)

For further information, see Council for Construction Robot Research (1999, pp. 234–235).

System 63: Mobile Concrete Exterior Floor Finishing Robot

Developer: Shimizu Corporation

The robot system is in form and functionality similar to Shimizu's Flat-KN but larger and was designed for the finishing of very large concrete surface areas in indoor and outdoor environments (Figure 2.141). It consists of three sets of trowel end-effectors located around a mobile robotic platform (depending on the finishing stage and task to be conducted, different types of trowel blades or discs can be installed to the end-effector), a control system that allows the robot either to be manually controlled or remotely operated, and an anti-collision system. The trowels are attached on the end of each retractable arm structure. The blade pressure against the floor can be adjusted in response to the hardness of the surface to be finished.

System 64: Mobile Concrete Floor Levelling and Screeding Robot

Developer: Takenaka Corporation

The Takenaka floor screeding robot (Figures 2.142 and 2.143) is used for the levelling and screeding of poured concrete. The robot consists of a lightweight and frame-like mobile platform, a set of concrete levelling and screening end-effectors mounted to the frame, and a quickly exchangeable on-board power supply unit. The mobile platform is able to move smoothly on very freshly poured concrete on four wire-cage–like wheels. The system was experimental and the frame structure of the mobile platform was designed to be able to host other subsystems and end-effectors that may be developed in the future.

Figure 2.142. Concrete floor smoothing robot, front view.

Figure 2.143. Concrete floor smoothing robot, side view.

Figure 2.144. Mobile floor finishing robot. (Image: Kajima Corporation.)

System 65: Mobile Floor Finishing Robot

Developer: Kajima Corporation

The robot was designed to automate the final labour-intensive stages of finishing concrete slabs (Figure 2.144). The robot consists of a mobile platform (equipped with a set of active drive wheels and passive steering wheels) that is capable of an almost omnidirectional movement, a main body equipped with tanks that can hold water or other substances used to assist the smoothing process as well as battery packs, a manipulator with 3 DOF (consisting of two rotational and one prismatic joint), and a trowelling end-effector with blade angle control activators. The mobile platform is equipped with a gyrocompass and linear distance sensor that allow the robot to navigate autonomously within a predefined work area. Work initiation and termination (including emergency stops) can be controlled remotely. Safety features of the robot include sensors for obstacle detection. The goal in the design of the robot was to increase the finishing quality (no more than 7 mm height difference within 3 m, surface roughness: less than 0.1 mm height difference within 30 cm of floor length), increase productivity (be able to finish up to 150 m^2 per man per day by using the robot), and increase the concrete finishing speed (be faster than 300 m^2, with the floor finishing speed synchronized with potentially possible concrete supply rates).

For further information, see Cousineau and Miura (1998).

2.11 Site Logistics Robots

Logistics operations on conventional construction sites, in particular those where many low-level parts must be handled, are numerous, time consuming, and involve hard physical work. Site logistics involve the identification, transportation, storage, and transfer (from one system or machine to another) of materials. Logistics operations follow at least along the main logistics routes on a construction site often clearly defined paths, and the interaction of a logistic system with the often diverse materials can in many cases be standardized through the use of pallets and containers. In particular, in the Japanese construction industry currently the automation of on-site logistics processes is becoming a standard event on conventional construction sites. This category includes systems for the automation of vertical delivery of material, systems that allow a horizontal delivery of material on the ground or on individual floors, systems that assist with the transfer of pallets or material (e.g., from a lift to a mobile platform), as well as automated material storage solutions. Systems for horizontal delivery of material can comprise state forklift-like mobile robotic platforms, rail-guided ground-based or ceiling-mounted systems, or smaller micro-/minilogistics solutions. In all these instances, the control and navigation complexity is reduced significantly by operating the robots along fixed, predefined routes with standardized interaction points with other systems. Payload capacities of the individual systems range from around 100 kilograms for micro-/minilogistics solutions to several hundred kilograms for systems for horizontal delivery to several thousand kilograms for vertical lifts. Operation speeds are situated in a range between 40 m/minute and 100 m/minute.

Site Logistics Robots		
66	Automated construction lift	Obayashi
67	Automated vertical material delivery lift	Kajima
68	Robotic construction lift	Cho et al., 2009
69	Automated guided vehicle (AGV)-based integrated on-site logistics system	Shimizu
70	AGV for construction sites	Kajima
71	Automated on-site delivery system	Obayashi
72	Automated on-site logistics system	Maeda and Komatsu
73	Automated horizontal/vertical on-site logistics system	Takenaka
74	Rail-guided overhead logistics system	Fujita
75	On-site monorail overhead logistics system	Kajima
76	Rail-guided automatic on-ground logistics	Kajima
77	Modular multirobotic on-site delivery and working platform system	Fujita
78	Automated ini logistics unit	Takenaka

System 66: Automated Construction Lift

Developer: Obayashi Corporation

This automated vertical logistics solution can be used as a separate system on the construction site, as well as in combination with other STCRs (e.g., robotic fork lifts,

Figure 2.145. Automatic vertical lift.

automated guided vehicles [AGVs], etc.), creating a larger material supply solution on the site (Figure 2.145). Obayashi also used the system in the context of its automated/robotic on-site factories. The system consists of two masts that allow lifting and lowering of the logistics compartment. The guiding masts are fixed to the ground as well as at certain heights to the building. The system allows the supply of small as well as large material (e.g., columns, beams) and construction equipment (including other STCRs). The logistics compartment is loaded on the ground level from material delivered from the storage area or by delivery truck by means of forklifts, cranes, or human workers. Following this, the logistics compartment lifts the components to the required floor level, where it can be unloaded. If required, two or more systems can be operated in parallel to guarantee an uninterrupted material supply.

System 67: Automated Vertical Material Delivery Lift

Developer: Kajima Corporation

The system was developed with the intention to arrange vertical transportation of goods on a construction site without the necessity of tower cranes (Figure 2.146).

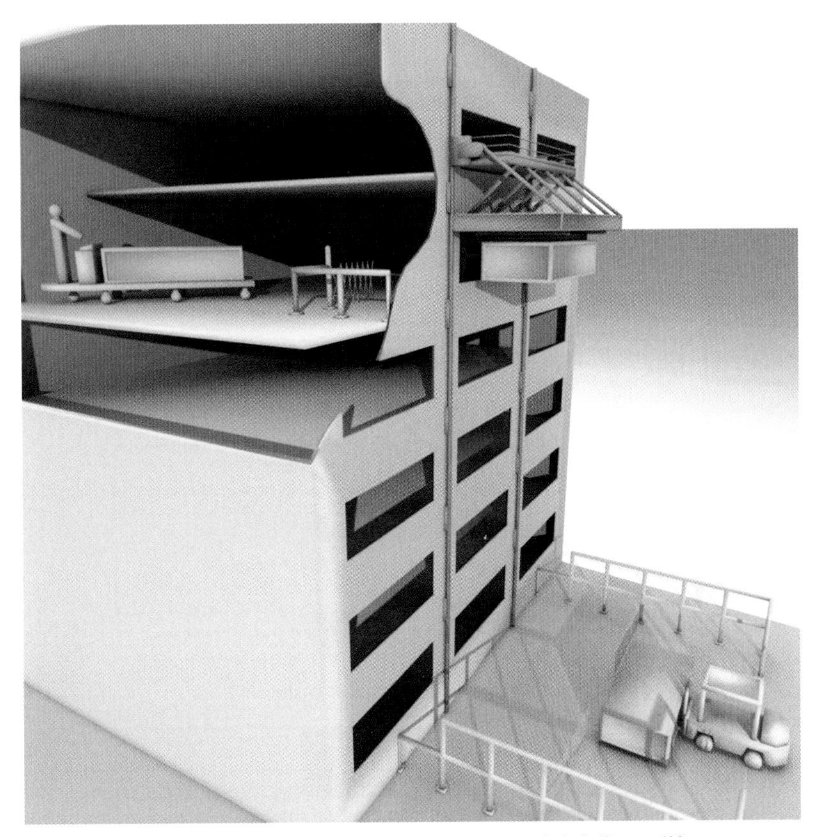

Figure 2.146. Overview of automated vertical material delivery lift system.

It is arranged so as to provide a continuous flow of construction materials, tools, and other elements. The system consists of a set of rails mounted to the outside of the building on which a logistics unit is operated in vertical direction. The logistics unit allows one to pick up standardized containers and compartments (placed under it or directly from delivery trucks or forklifts) from atop. It then delivers these containers to the floor where the material is required and transfers it through a cantilevering, telescopic transfer mechanism into the floor onto a delivery template. The system can also be used in renovation and building disassembly/deconstruction projects.

System 68 Robotic Construction Lift

Developer: Dr. Cho and Dr. Kwon (SKK University, Republic of Korea)

Construction lifting tasks require the vertical transfer of construction materials and delivering of those materials to the desired location. In particular, in a high-rise building construction project, an efficient on-site logistic management plan is vital and can potentially determine the success of the overall project. To ensure effective on-site logistics operations, it is critical to optimize the lift operating plan and assist the construction managers and workers with the decision-making process. The robotic construction lift is an attempt to develop a hybrid intelligent lifting, planning, and management system that allows optimizing on-site logistics operations. The lift system consists not only of the mechanically active subsystems (an intelligent lift,

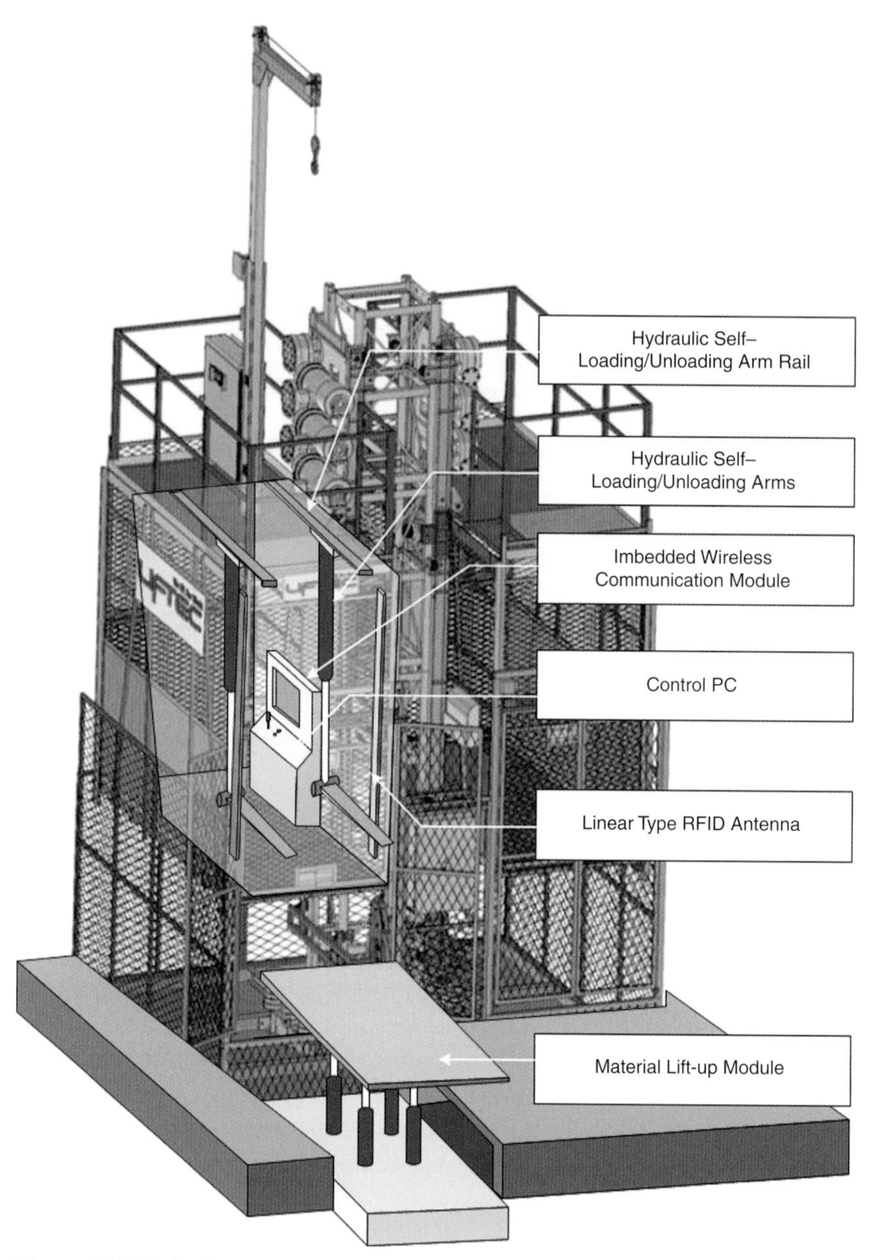

Figure 2.147. Robotic construction lift. (Image: Cho et al., 2009.)

material lift-up module; Figure 2.147) but also an informationally active logistics procedures optimization system. The intelligent lift is composed of a loading cage and a hydraulically activated self-loading/unloading system. The material lift-up module functions as an intermediary between other systems that deliver material horizontally to the lift and allows this material to be brought into a position that allows the forks of the hydraulically activated self-loading/unloading system to pick it up. First, automated recognition of the destination floor for each material loaded into the lift is carried out with the help of a project management information system (PMIS).

Figure 2.148 AGV transporting material from the automated construction lift to the assembly area on an individual floor.

The radio frequency identification tag (RFID)-scanning module (composed of linear type RFID antennas) identifies the desired materials for each destination floor and identifies the material.

For further information, see Cho et al. (2009).

System 69: AGV-Based Integrated On-Site Logistics System

Developer: Shimizu Corporation

This on-site logistics system was designed for the delivery of construction materials such as plasterboards, fittings, pipes, ventilation components, and electrical components (Figure 2.148). The system consists of an automated vertical construction lift; an AGV; a control and scheduling system; and fixed, marked travelling routes (Figure 2.149). The AGV utilizes a mobile platform equipped with a fork lifting mechanism (capacity of 1300 kg) that allows it to take up and unload materials. It is equipped with an on-board battery power supply (it can be recharged at dedicated recharging locations to which it can automatically move) and can move only along fixed, marked trajectories, which significantly reduces the control and sensor complexity (i.e., no sophisticated sensors for navigation are necessary). As the operations of the lift and the AGV can be integrated and synchronized, it can be used on the ground floor to automate the delivery of material from a storage area into the lift and then to the individual floor level as well as the delivery of materials on floor levels from the lift to the assembly area. The vertical lift allows the AGV to move onto it for loading/unloading material or pallets. The system is thus able to fully automate the logistics on the site along predefined logistics routes. Further specifications of the system are given in Table 2.4.

Table 2.4. *AGV-Based Integrated On-Site Logistics System Specifications*

AGV	
Type	Side-fork type lifting AGV
Rated capacity	1300 kg
Rated speed	40 m/min
Radius of turn	3
Dimensions	L = 2.50 m, W = 1.30 m, H = 1.00 m
Dry weight	700 kg
Power source	DC24V battery
Automated vertical construction lift	
Rated capacity	2000 kg
Rated speed	100 m/minute
Lift height	Max. 300 m
Power source	3 AC200V, 50/60 Hz

System 70: AGV for Construction Sites

Developer: Kajima Corporation

This forklift type AGV was developed by Kajima to automate the delivery of material along predefined, fixed logistics routes (Figure 2.150). It consists of a robotic mobile platform integrated with a forklift mechanism. Similar to other AGVs or robotic forklifts designed for construction sites, it can be used on the ground floor to automate the delivery of material from a storage area to a vertical construction lift as well as to automate the delivery of materials on individual floor levels from

Figure 2.149. The AGV can move only along fixed, marked trajectories what significantly reduces control and sensor complexity.

Figure 2.150. AGV for construction sites system in action.

a vertical construction lift to the assembly area. The system can be informationally integrated with other automated equipment and STCRs on the site to build up larger material handling and material supply chains.

System 71: Automated On-Site Delivery System

Developer: Obayashi Corporation

The system was designed to reduce delivery costs and in particular enhance the efficiency of high-rise construction. The system consists of four main parts: (1) an automatic transfer equipment, (2) an automated guided forklift (AGF), (3) an automated storage rack, and (4) a web-based delivery scheduling system (WSS). The automatic transfer equipment is an intermediary system (capacity: 200 kgf) that can unload materials from a vertical construction lift to a dedicated location in front of the lift where the AGF can the take up the material (Figure 2.151). Instead of using on-board battery power, the power for the automatic transfer equipment is provided through a flexible cable connected to the vertical construction lift's power source, thus minimizing its weight (800 kgf). The AGF has a lifting capacity of 1500 kgf, and has a compact chassis to be able to access all areas on the site. It runs on electromagnetically marked, fixed trajectories that connect dedicated load/unload area in front of the lift with other areas on the floor (e.g., the automated storage rack) where the AGF operates. The automated storage rack that is part of the system serves as an intelligent warehouse that operates without human interaction and allows for direct and automated interaction with the AGF (Figure 2.152). Deployed on the ground

Figure 2.151. The automatic transfer equipment is an intermediary system that can both take up material from a dedicated location in front of a vertical lift to which the AGF has delivered material and unload a vertical lift. (Image: Obayashi Corporation.)

Figure 2.152. The AGF can directly interact with an automated storage system.

Figure 2.153. Material is transported by the vertical lift to a certain floor and moved through a transfer mechanism that is part of the lift onto the elevated track in front of the lift. (Image: Maeda Corporation.)

level the AGF can connect the automated storage rack with the automatic transfer equipment/vertical lift and thus automate the supply of material from the storage to an individual floor where the material is required. Deployed on an individual floor the AGF can automate the delivery of the material from the lift to the area where the material should be processed or assembled. The WSS consists of a terminal for application (TA), a web-shared server (WS), and a terminal for management (TM). The main purpose of the TM is data collection/input and delivery scheduling. These data are then uploaded to the WS so that the information is accessible for human workers and the automated equipment.

For further information, see Atsuhiro et al. (2002).

System 72: Automated On-Site Logistics System

Developer: Maeda Corporation and Komatsu Ltd.

The system was developed jointly by Maeda and Komatsu and consists of an automated vertical construction lift, a robotic forklift, and an elevated track in front of the lift (Figure 2.153) that serves as an intermediary between the automated lift and the robotic forklift. Material is transported by the vertical lift to a certain floor and moved through a transfer mechanism that is part of the lift onto the elevated track in front of the lift. The elevated track thus serves as a standardized interaction template between the vertical lift and the robotic forklift. The robotic forklift

Figure 2.154. The robotic forklift can pass with its forks under the elevated track and thus lift the pallets placed on it. (Image: Maeda Corporation.)

can pass with its forks under the elevated track and thus lift the pallets placed on it (Figure 2.154). It can thus take up the material and move it to the required location on the floor. This process works also in the opposite direction and the robotic forklift can place pallets on the elevated track from its location near the lift and automatically transfer it to the lift.

System 73: Automated Horizontal/Vertical On-Site Logistics System

Developer: Takenaka Corporation

Takenaka's automated horizontal/vertical on-site logistics system is able to automate logistics operations along the main logistics routes on construction sites. It consists of an automated storage system integrated with the system, several robotic forklifts, and an automated vertical material delivery lift (Figures 2.155 and 2.156). The robotic forklifts pick up the material stored in standardized and system compatible pallets of boxes and deliver it to and load it into the automated vertical material delivery lift. The lift delivers the material to the desired floor, where other robotic forklifts or human workers pick up the material. The robotic forklifts, the key element of the system, consist of a mobile platform able to navigate autonomously on preprogrammed routes and forklift end-effectors. The anti-collision sensing system helps the robot to avoid obstacles in its travel path. The on-board command module and communication system provide an interface between the robot and other on-site equipment.

Figure 2.155. Robotic horizontal on-site forklift. (Image: Takenaka Corporation.)

System 74: Rail-Guided Overhead Logistics System

Developer: Fujita Corporation

This overhead crane logistic system developed by Fujita is used to transport building materials horizontally on individual floors (Figure 2.157). The system consists of rails installed along the intended logistics paths along which a logistics unit can then move. The rails are installed towards the ceilings of already constructed floors or any dedicated temporary structure serving this purpose. The logistics unit contains both a mechanism to lift the material up/down through a hoist as well as to move along the rails to the desired locations. The building material to be transported is tied manually to the hoist of the crane. The lifting of the material by the logistics unit as well as the movement along the rails can be conducted automatically or in a remote-controlled manner.

Figure 2.156. Interaction of robotic horizontal on-site forklift with automated vertical lift. (Image: Takenaka Corporation.)

Figure 2.157. Rail-guided overhead logistics system. (Image: Fujita Corporation.)

Figure 2.158. On-site monorail overhead logistics system. (Image: Kajima Corporation.)

Figure 2.159. Rail-guided automatic on-ground logistics. (Image: Kajima Corporation.)

System 75: On-Site Monorail Overhead Logistics System

Developer: Kajima Corporation

This overhead crane logistic system developed by Kajima is used to transport building materials horizontally on individual floors (Figure 2.158). The system consists of rails installed towards the ceiling along the intended logistics paths along which a logistics unit then can move. The rails are installed towards the ceilings of already constructed floors or any dedicated temporary structure serving this purpose. The logistics unit is equipped with two end-effectors for taking up/lifting material. They can be used both to carry two spate loads or to carry jointly one larger load. Once the material is lifted, the logistics unit can be driven with the material suspended from it along the overhead rails to the desired location.

System 76: Rail-Guided Automatic On-Ground Logistics

Developer: Kajima Corporation

Kajima's rail guided on-ground logistics system (Figure 2.159) can be used to automate the transport of material from the material delivery yard or on-site factory

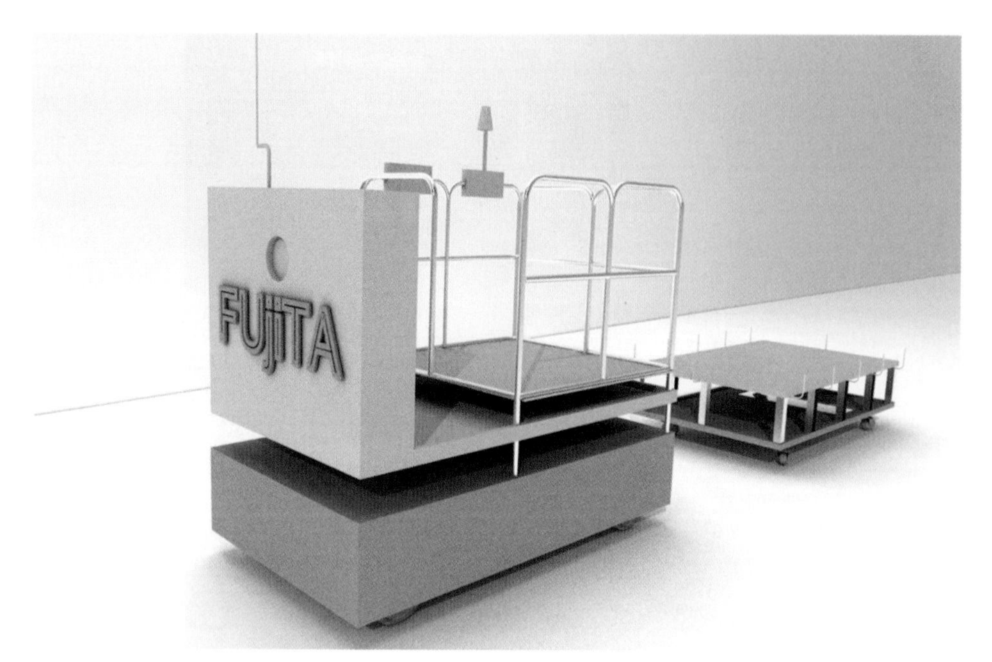

Figure 2.160. Modular multirobotic on-site delivery and working platform system, Fujita Corporation.

to the system that lifts the material from the ground to the floor where it is to be installed. Its rail-guided nature makes it suitable, in particular, for automating the main transportation routes in the construction site. The system consists of a logistics frame, a rail system, and a motor unit that pulls the logistics frame along the rail and is placed at one end of the rail system. The system's efficiency is maximized when logistics routes on site are concentrated and standardized. In the event that the logistics system is connected to a just in time delivery of material on one end and to a fully integrated and/or automated component lifting system (e.g., a vertical delivery system of an automated/robotic on-site factory) on its other end, a production line–like material flow can be established. The guiding rails make the system comparatively complex to set up, but on the other hand, this method allows the system itself to be relatively simple and robust.

System 77: Modular Multirobotic On-Site Delivery and Working Platform System

Developer: Fujita Corporation

Fujita's modular robotic on-site delivery and working platform (Figure 2.160) system can automate the transportation of material on the construction site and it can be used to lift into place material, components, and human workers, thus rendering the comparably labour-intensive installation of temporary scaffolding obsolete. The system is relatively lightweight and can be used on ground as well as on above-ground floor levels. The system is also not bound to guiding rails or other guiding systems and can therefore reach any distinct location on a dedicated floor. The system consists of a mobile robotic platform, a transport and lifting platform installed on top of the mobile robotic platform, and a separate set of attachable mobile material pallets.

Figure 2.161. Automated mini logistics unit. (Image: Takenaka Corporation.)

System 78: Automated Mini Logistics Unit

Developer: Takenaka Corporation

The system was developed to automate on the construction site the transport of light and smaller sized materials on individual floors. The system consists of a mobile robotic platform to which different types of movable pallets can be connected (Figure 2.161). The pallets can be manually loaded with material (e.g., pipes or fittings) and then transported by the mobile platform to preprogrammed destinations.

2.12 Aerial Robots for Building Structure Assembly

In the past decade, a vast number of aerial robots of different sizes and types were developed for both civil and military use. In the civil field aerial drones are today no longer considered to be useful only in the context of tasks such as inspection, surveillance, or surveying, as first attempts are currently under way to use drones as logistics systems in manufacturing in factories as well as outdoors. In the military field drones capable of carrying heavy payloads and with considerable ranges have become key equipment. In light of the fast pace of innovation in the field of aerial robots and inspired by the theoretical potentials that aerial robots might hold for construction, researchers are about to explore the concepts and approaches that would allow the utilization of the potentials of arterial robots (beyond surveying and monitoring capabilities, which are outlined in Section 2.1) for site logistics and building structure assembly. However, despite the obvious advantages of such a use of arterial robots in construction (e.g., independence from roads and other infrastructures, freeing the site from heavy equipment such as cranes, etc.) the challenges with respect to a full-scale implementation are enormous (payloads, power supply, assembly approaches, flight stabilization strategies, etc.). Researchers, therefore, are currently focussing initial efforts on experimental test beds and scale models of aerial assembly and logistics scenarios to explore relevant methods, technologies, flight trajectories, algorithms,

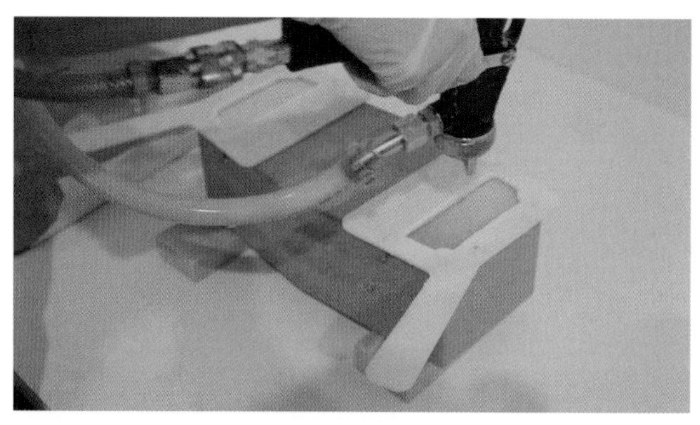

Figure 2.162. Preinstallation of adhesion stripes. (Image: Flight Assembled Architecture © Gramazio & Kohler and Raffaello D'Andrea in cooperation with ETH Zurich at the FRAC Orléans, co-producer.)

building modularity, component connectors, assembly sequences, and automation pipelines.

Aerial Robots for Building Structure Assembly		
79	Flight-assembled architecture	Gramazio & Kohler and Raffaello D' Andrea in cooperation with ETH Zürich at the FRAC Orléans
80	Aerial construction with qaudrotor teams	GRASP Lab, University of Pennsylvania

System 79: Flight-Assembled Architecture

Developer: Gramazio & Kohler and Raffaello D'Andrea, in cooperation with ETH Zurich at the FRAC Orléans (co-producer)

The system was developed and demonstrated (at the FRAC Orléans) as an experimental (scale model type) mock-up for exploring the feasibility of the erection of large high-rise buildings by aerial vehicles and to explore possible strategies and approaches. In the experiment, a set of quadrocopters picks lightweight building blocks (polystyrene foam bricks) one by one from a dedicated picking table presenting the building blocks in a standardized manner (and with adhesion stripes preinstalled on their bottom side) to the quadrocopters (Figures 2.162 and 2.163). Using the building blocks, the quadrocopters build a predefined tower structure (Figures 2.164–2.166). In the future, developers plan to further integrate and automate the whole process from planning to construction to ultimately create a parametric system that is able to generate the detailed block structure as well as the flight paths of the quadrocopters based on an input geometry of the desired building. The experiment at FRAC Orléans was conducted in an indoor environment fostering the accurate control of quadrocopters and flight and positioning processes through a combination of local on-board sensors and a global vision-based sensing system. The ETH Zurich research team sees flying robots cooperating as swarms and with other robots as an important means for logistics and assembly in the context of the construction of future megacities and megastructures.

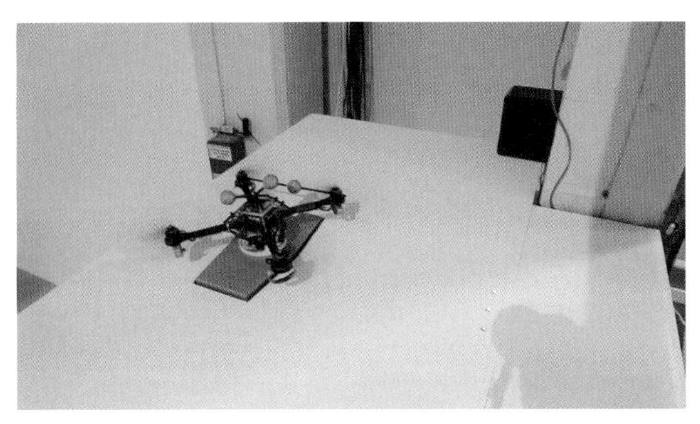

Figure 2.163. Quadrocopters pick the building blocks from a picking table presenting the building blocks in a standardized manner to the robot. (Image: Flight Assembled Architecture © Gramazio & Kohler and Raffaello D'Andrea in cooperation with ETH Zurich at the FRAC Orléans, co-producer.)

Figure 2.164. Detailed view of a quadrocopter. (Image: Flight-assembled architecture. © Gramazio & Kohler and Raffaello D'Andrea in cooperation with ETH Zurich at the FRAC Orléans, co-producer.)

Figure 2.165. Quadrocopter launching and recharging pads. (Image: Flight Assembled Architecture © Gramazio & Kohler and Raffaello D'Andrea in cooperation with ETH Zurich at the FRAC Orléans, co-producer.)

Figure 2.166. Assembly of a scale model of a megastructure by a swarm of quadrocopters. (Image: Flight-assembled architecture © Gramazio & Kohler and Raffaello D'Andrea in cooperation with ETH Zurich at the FRAC Orléans, co-producer; Image by Francois Lauginie.)

System 80: Aerial Construction with Quadrotor Teams

Developer: GRASP Lab, University of Pennsylvania/Q. Lindsey, D. Mellinger, V. Kumar

The system makes it possible to quickly and autonomously build up tower-like structures by a swarm of aerial robots (quadrocopters). One objective was to create an automation pipeline in which cubic structures are generated from defined geometries, assembly procedures are derived, and finally the cubic structures (special cubic structures [SCS]) accordingly are built from specially designed, robot-adapted building blocks by quadrocopters (Lindsey et al. 2012). The quadrocopters (type: Hummingbird from Ascending Technologies; weight: 500 grams; operation time: 20 minutes; payload: 500 grams) are equipped with a gripper that allows picking up the building blocks (representing columns and beams of a building's structure) in a vertical or horizontal position. The gripper was designed to be fully in tune and complementary with the building blocks and the connectors (Figures 2.168 and 2.169). A high construction speed and work parallelization were achieved through use of multiple copters (Figure 2.170). The site is separated into a picking area (where the building blocks are presented to and fed to the aerial robots accurately and in the correct positions and sequence) and a construction area (Figure 2.167). Connector systems installed at the ends of the building blocks allow the blocks to be fixed to each other. The connector system uses magnets not only to fix the building block but also to fine adjust and align their position once brought roughly to the desired location. This mechanism potentially also allows aerial robots providing a lower positioning accuracy to be used. Utilizing a yawing moment the aerial robots can identify if building

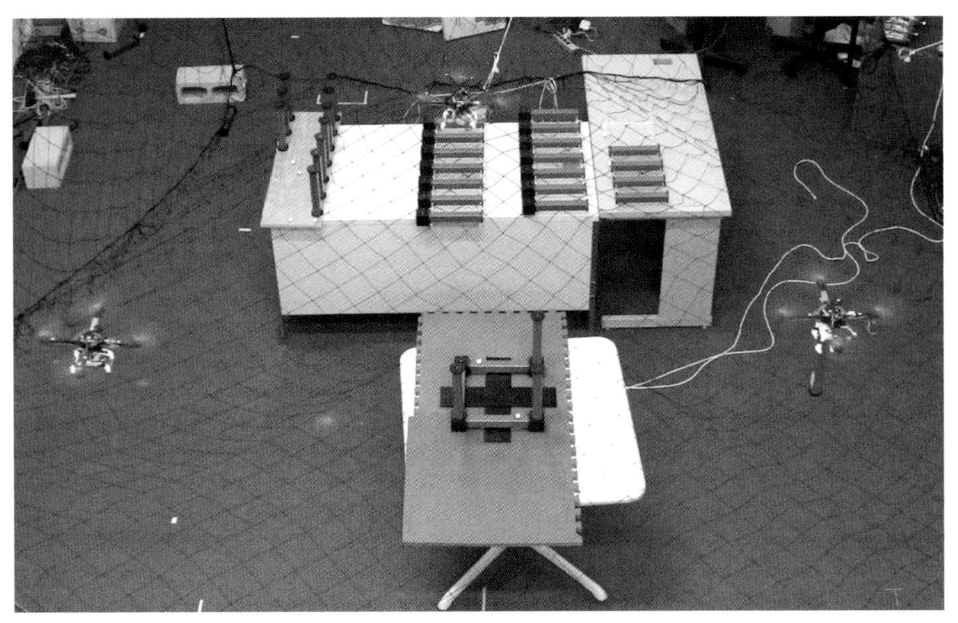

Figure 2.167. The site is separated into a picking area (where the building blocks are presented to aerial robots accurately and in the correct positions and sequence) and a construction area. (Image: GRASP Lab/Lindsey, Mellinger, Kumar.)

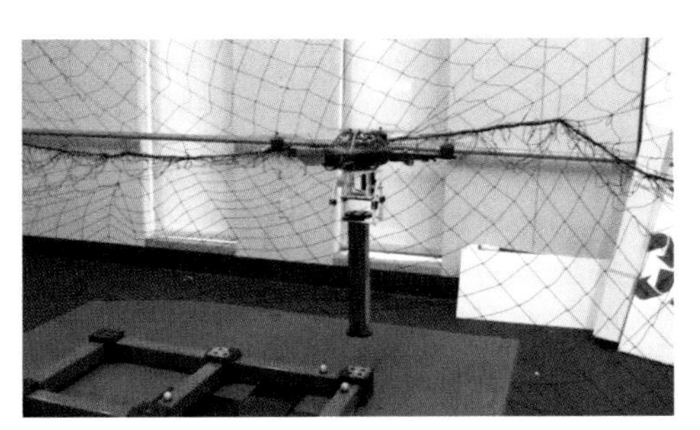

Figure 2.168. Quadrotors are equipped with a gripper that allows picking up the building blocks in vertical or horizontal position. (Image: GRASP Lab/Lindsey, Mellinger, Kumar.)

Figure 2.169. Quadrotor with part. (From Lindsey et al., 2012, "Construction with quadrotor teams." *Autonomous Robots*, 33: 323–336, with permission of Springer Science+Business Media.)

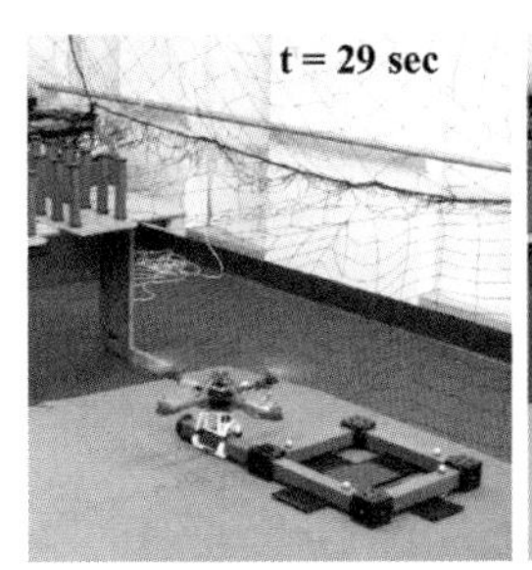

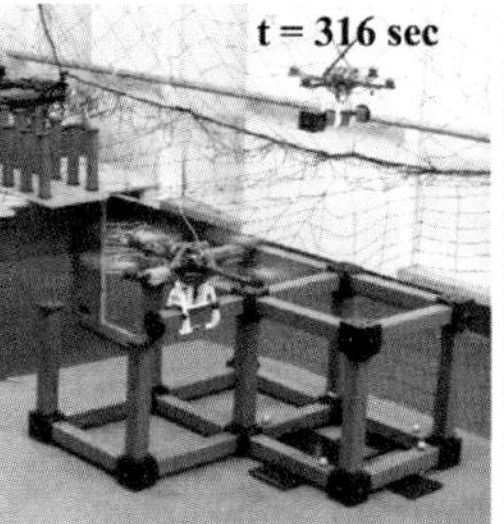

Figure 2.170. Intermediate snapshots of a pyramid-like SCS being built by three quadrotors. (From Lindsey et al., 2012, "Construction with quadrotor teams," *Autonomous Robots*, 33:323–336, with permission of Springer Science+Business Media.)

Figure 2.171. Both robots and building block also were developed further to be able to assemble tetrahedal structures. (Image: GRASP Lab/Lindsey, Mellinger, Kumar.)

blocks have been placed successfully. The autonomous construction of a cubic structure was demonstrated as a scale model in an indoor environment. Both robots and building block also were developed further to be able to assemble tetrahedal structures (Figure 2.171).

For further information, see GRASP Lab (2015) and Lindsey et al. (2012).

2.13 Swarm Robotics and Self-Assembling Building Structures

Robotic swarm approaches build on multirobot collaboration and the interaction of a multitude of usually relatively simple and standardized individual robotic building blocks. Many approaches build on advanced distributed control architectures, consider ideas of self-regulation and organization of systems, and are inspired by the collective organization, behaviour, and communication of insects such as termites, ants, or bees. In particular, in the context of construction where the "product" in most cases indeed is assembled from a multitude of similar, standardized building blocks or elements this seems to be logical approach. However, the challenges (e.g., scale of building projects, real-world site conditions, etc.) such a system would face in full-scale deployment are still subjects of ongoing research, and approaches in the context of the use of swarms of robots in construction can so far be allocated predominately in the field of experimental and scale-model–based research.

In the context of construction, the idea of multirobot systems that can climb in and assemble structures in the form of a stand-alone system or as swarms has its roots in space construction research done in the 1980s and 1990s. The developed systems focussed mainly on the assembly of truss-based space structures by assembly robots that can climb and move within these structures (e.g., Shimizu Space Walker/Yoshida & Ueno 1990; Skyworker Project/Skaff et al. 2001; NASA/Doggett 2002). With the advance of the research field, more complex scenarios for construction on earth (which requires significantly more sophisticated robots owing to the presence of gravity and a potentially larger variety of structures) were tackled.

Two basic research directions formed over time. One direction builds on swarms of robots that can climb within structures and manipulate (swarms of) components. The components can be either simple structural elements (e.g., Nigl et al. 2013) or mechatronic/robotic components (e.g., Terada & Murata 2008; Werfel 2012). The other research direction builds on swarms of robots that are both assembly system and component (e.g., Yim et al. 2003; Molecubes 2015) in one entity. The first approach allows the complexity to be distributed, and, in particular, the components to be assembled are simple and suited for mass production. The second approach has the advantage that only one element has to be designed and coordinated, but critics remark that with the element in place in the structure, the capability of the robot cannot be used actively, and the robot rests, resulting in a kind of idle time.

Beyond the mechanical robot systems and hardware, Terrada and Murata (2008) and Werfl (2012) have developed parametric software that allows one to initiate the construction process by simply specifying the shape of the desired structure to be built. Furthermore, open source strategies were tested that allow a case-based modification and customization of robots or robot components, for example, through 3D printing of robot parts (Zykov et al. 2008).

Swarm Robotics and Self-Assembling Building Structures		
81	Automatic modular assembly system (AMAS)	Y. Terada and S. Murata, Tokyo Institute of Technology
82	TERMES – termite-inspired robot construction team	Harvard University's Wyss Institute for Biologically Inspired Engineering
83	Molecubes	Viktor Zykov and Hod Lipson
84	Structure-reconfiguring robots	Franz Nigl, Shuguang Li, Jeremy E. Blum, and Hod Lipson
85	PolyBot	Palo Alto Research Center (PARC), Mark Yim
86	Space construction robotics technology	Carnegie Mellon University, Shimizu Corporation

System 81: Space Construction Robotics Technology

Developer: Carnegie Mellon University, Shimizu Corporation

Shimizu is a Japanese construction firm that is actively seeking new applications for construction services. One of the fields the construction firm has been actively exploring for decades is space construction. Shimizu therefore set up a dedicated department researching construction in space. This department is closely connected also to Shimizu's construction automation and robotics research group. At the beginning of the 1990s, Shimizu developed jointly with Carnegie Mellon University (at that time and also today one of the leading institutions in robotics research) a series of concepts and prototypes of robots that can climb in/on structures and

Figure 2.172. Testing of movement for walking over the built structure at Carnegie Mellon University. (Image: Tetsuji Yoshida, Shimizu Corporation.)

Figure 2.173. Detail of end-effector used for climbing over the built structure at Carnegie Mellon University. (Image: Tetsuji Yoshida, Shimizu Corporation.)

assemble, reconfigure, and maintain them. The "Spacewalker" concept, for example, aimed at rather simple robots with only two or three main links that could be used in a swarm to move over a truss-like structure and assemble or reconfigure it. Key components of the robots were prototyped and lab-tested (Figures 2.172 and 2.173). With the increasing use of lightweight and systemized steel structures in construction (facades, building systems, etc.) such concepts become increasingly interesting for building construction on earth.

System 82: PolyBot

Developer: Palo Alto Research Center (PARC), Mark Yim

The PolyBot is a modular multirobot system composed of individual robotic segments and nodes. These robotic segments and nodes can attach to and detach from each other to form a variety of forms such as snakes-, loop-, or spider-like configurations for certain mobility and manipulation purposes. The system was intended as a robot system for conducting construction servicing and maintenance tasks in space as well as a robot system to be used in tasks such as search and rescue (e.g., in wheel mode [Figure 2.175] or spider-like configuration [Figure 2.176]) and providing shelter for victims of disasters (e.g., by transforming into a dome). The ability of self-reconfiguration would allow a variety of manipulation tasks in different use cases to be accomplished by the same system, erasing the need for different, specialized robots to be sent into space. The robotic segments come in different configurations and consist basically of one activated rotational joint (Figure 2.174). The hermaphroditic connectors, in particular of PolyBot version 3, allow for self-reconfiguration and consist of a set of pins, chamfered holes, electrical connectors, and infrared docking guidance systems to physically, informationally, and in terms of power supply join, detach, and rejoin the individual robotic segments. The prototyped robotic segments of version 3 had a size of W 5 cm × L 5 cm × H 4.5 cm.

For further information, see Yim et al. (2003).

Figure 2.174. G3 PolyBot module prototype. (From Yim et al., 2003, "Modular reconfigurable robots in space applications, *Autonomous Robots*, 14:225–237, with permission of Springer Science+Business Media.)

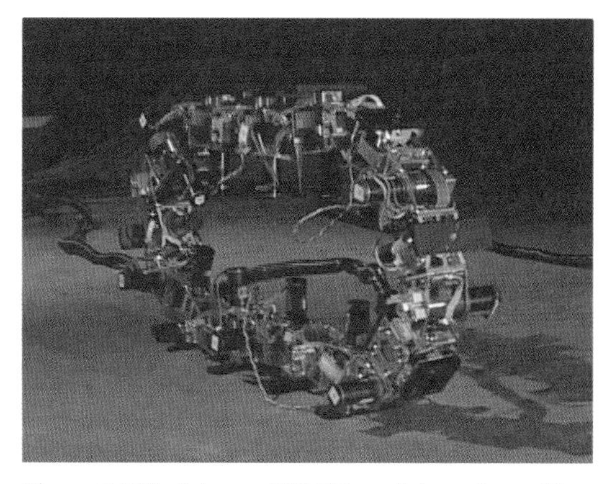

Figure 2.175. A loop of 23 G2 modules using rolling track locomotion. (From Yim et al., 2003, "Modular reconfigurable robots in space applications, *Autonomous Robots*, 14:225–237, with permission of Springer Science+Business Media.)

System 83: Automatic Modular Assembly System (AMAS)

Developer: Y. Terada and S. Murata, Tokyo Institute of Technology

The AMAS is a multirobot system that utilizes approaches from the field of distributed, swarm-like intelligence. Its key elements are standardized, robotic building

Figure 2.176. A four-legged spider-like configuration with G2 modules. (From Yim et al., 2003, "Modular reconfigurable robots in space applications," *Autonomous Robots*, 14:225–237, with permission of Springer Science+Business Media.)

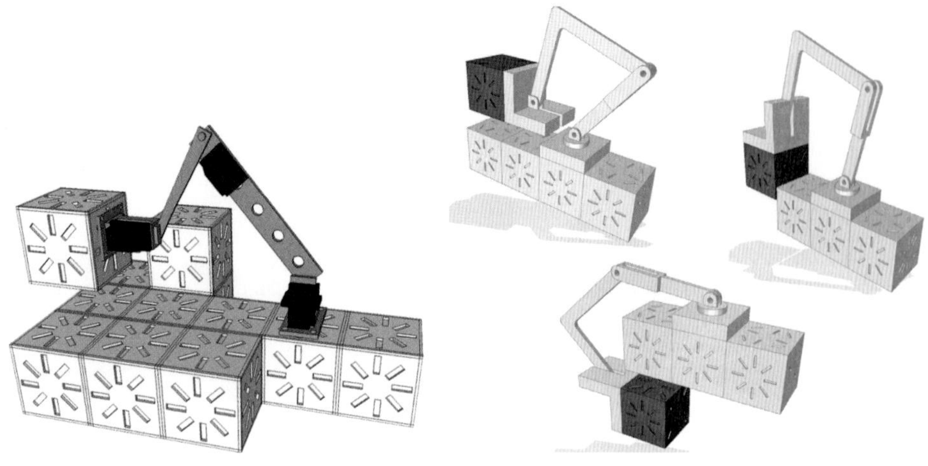

Figure 2.177. AMAS mock-up. (Image: Own interpretation according to Terada & Murata, 2008.)

Figure 2.178. Assembly capability. (Image: Own interpretation according to Terada & Murata, 2008.)

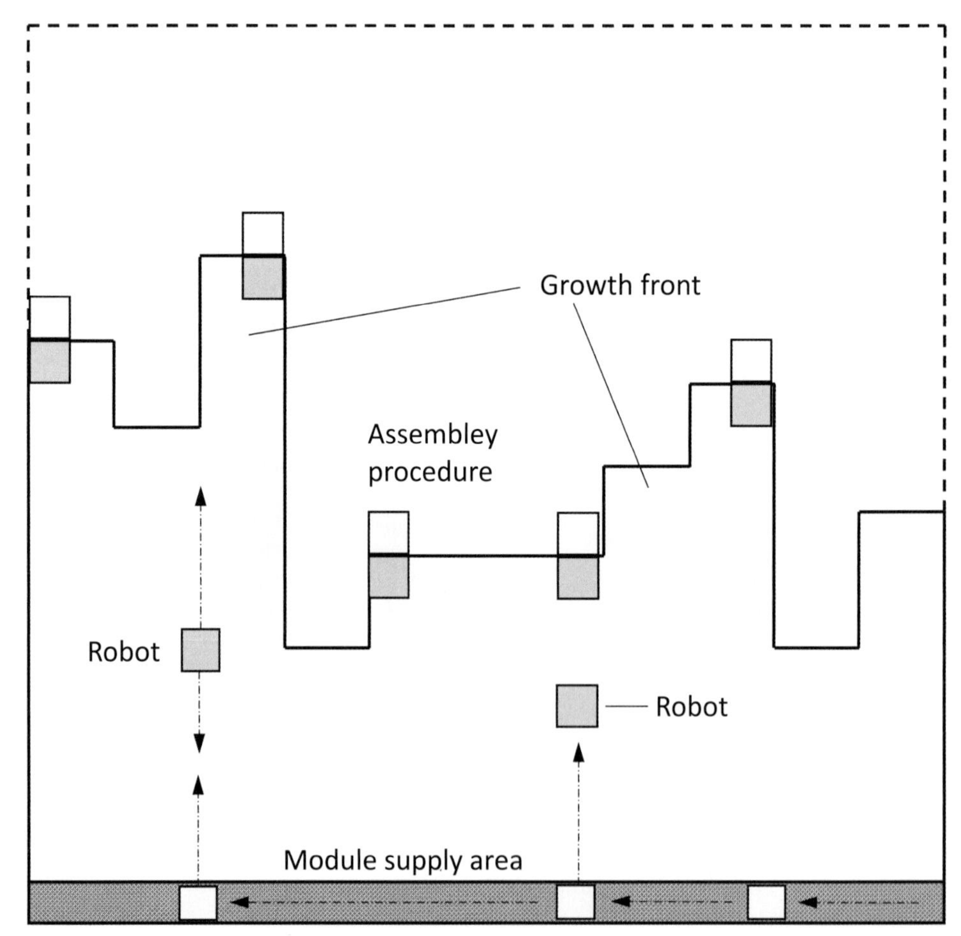

Figure 2.179. Overall material supply and assembly process on a construction site. (Image: Own interpretation according to Terada & Murata, 2008.)

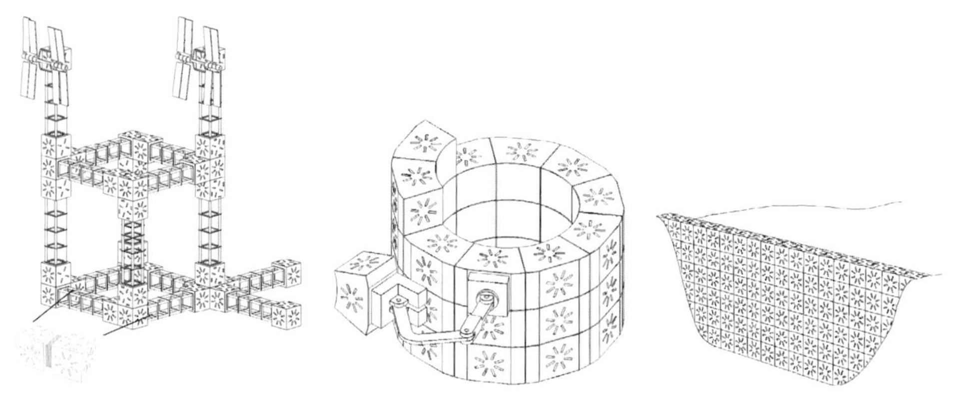

Figure 2.180. Future application scenarios. (Images adopted from Terada and Murata, 2008, "Automatic Modular Assembly System and its Distributed Control", *The International Journal of Robotics Research*, Vol. 27, No. 3–4, SAGE Publishing: March/April 2008, pp. 445–462, doi: 10.1177/0278364907085562, with permission from SAGE Publishing.)

blocks; assembler robots; and a distributed control architecture (Figures 2.177 and 2.178). The assembler robots consist of a robotic arm connected to two end-effectors, one on each end. Each of the end-effectors works differently: one (carrying base) can manipulate the building modules (pick them up, position, align, and fix them); and the second one (motion base) can couple with the assembled building blocks, securely fix the robot for manipulation tasks to the structure, and use the structure to climb and move in an inchworm-like way. The building blocks can also be considered as robotic elements and each one contains a microprocessor, sensors (infrared and contact sensors), and complex connectors. As the building blocks are thus intelligent as well and are fully integrated physically and in an informational sense with the assembler robots, both building block and assembler robots are components of the robotic system. A distributed gradient field algorithm is used to control the assembly activity of multiple assembler robots. The gradient filed is generated through inter-module communications allowing for a self-organization of the system. On the construction site the assembler robots pick up the building blocks from the module supply area, transport them to the growth front (where actual construction takes place), and assemble them (Figure 2.179). It is expected that AMAS can be used to construct large-scale structures such as space structures, bridge pylons, large pillars, and dams (Figure 2.180).

For further information, see Terada and Murata (2005) and Terada and Murata (2008).

System 84: TERMES – Termite-Inspired Robot Construction Team

Developer: Harvard University's Wyss Institute for Biologically Inspired Engineering

TERMES is a decentralized multirobot system (Figures 2.181 and 2.182) developed by Harvard University's Wyss Institute for Biologically Inspired Engineering building on the concept of swarm robotics. The approach follows the organization, behaviour, and communication of termites building their complex

Figure 2.181. The system consists of a set of simple and independent robots equipped with on-board microprocessors, sensors and activators, and passive building blocks with passive mechanical features. (Image: Eliza Grinnell, Harvard School of Engineering and Applied Sciences.)

Figure 2.182. The robots interact with their environment in real time, using a set of on-board sensors only and determine what to do according to what they encounter. (Image: Eliza Grinnell, Harvard School of Engineering and Applied Sciences.)

colonies without understanding on the individual level what the overall structure looks like. The system consists of a set of simple and independent robots equipped with on-board microprocessors, sensors and activators, and passive building blocks with passive mechanical features. The robots' sensors can be classified into four types: infrared, ultrasonic, accelerometer, and contact sensors. To provide position feedback, each robot is equipped with six active infrared sensors to sense the location of bricks and navigate its way around the structure. Each robot is equipped with a manipulation arm and an end-effector to be able to manipulate specially designed blocks with added passive mechanical features (for connection to other components or the robot's end-effector: chamfers for self-alignment and magnets for fixation) to help the robot grasp and securely align each block in its place in the intended structure. The robots move using four small curved whegs (wheel-legs) to move and climb. The robots do not communicate with the environment or the building blocks, have information on only a model of the desired structure, and inspect their environment ad hoc through continuous movement. An offline compiler is used to convert a desired building design (high level-representation of the target structure) into a "structurepath" representation that provides simple guidelines for the robots' movements. The robots interact with their environment in real time, using a set of on-board sensors only (no global sensors needed) and determine what to do according to what they encounter. Each robot acts independently and the more robots added to the process, the faster the project is completed, as the robots can work in parallel.

For further information, see Werfel et al. (2014).

System 85: Molecubes

Developer: Viktor Zykov and Hod Lipson

Molecubes are modular, compact robot cubes that can interact as a swarm to build up structures composed of the robot cubes themselves or to combine into robotic arm- or inchworm-like structures that can then assemble truss elements or other building structures by moving along these structures (Figures 2.183 and 2.184).

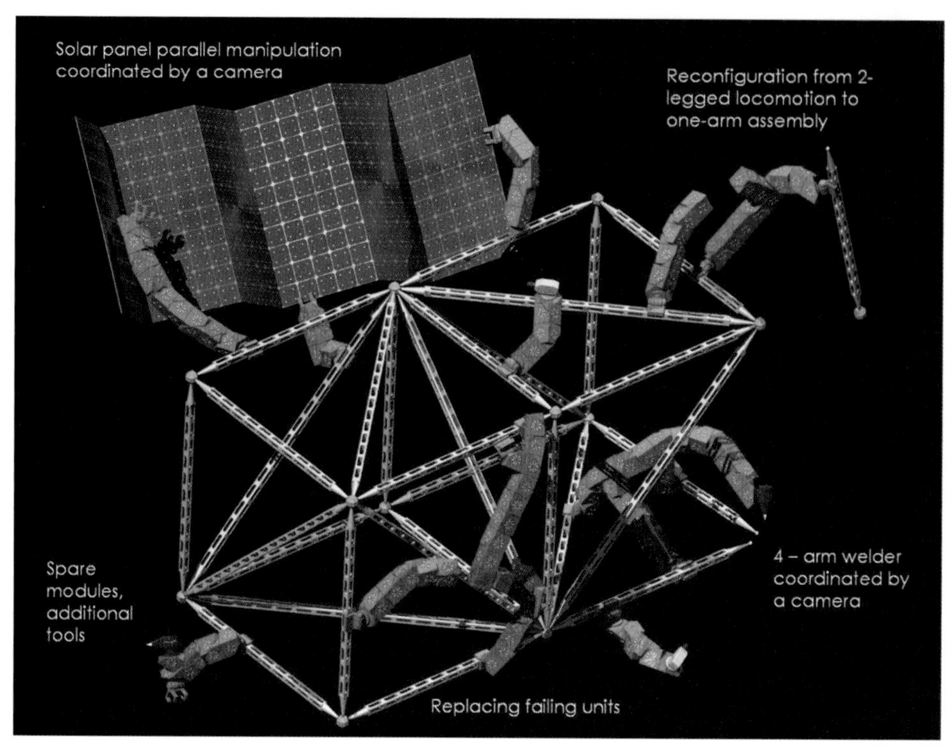

Figure 2.183. Molecubes interacting as a swarm and performing construction, self-repairing, coordinated manipulation, and reconfiguration activities. (Image: © [2007] IEEE. Reprinted, with permission, from Zykov et al. 2007.)

Figure 2.184. Molecubes combine into a four-unit robotic arm that can assemble a copy of itself. (Image: Viktor Zykov.)

The basic Molecube entity is a robotic cube that consists of two movable halves, each equipped with a connector, which allows it to join with another robotic cube. Each of the halves of the robotic cube can rotate independently, allowing for a variety of motions and configurations both as a single entity and a connected swarm. To facilitate fast and safe joining, a magnetic mechanism is part of the joining system. The system is available as a modular, scale model version for research purposes. Furthermore, recent approaches aim at making the system open source, allowing individual development teams to download, modify, and produce Molecubes themselves, for example, by 3D printing.

For further information, see Zykov et al. (2005, 2007) and Zykov et al. (2007).

System 86: Structure-Reconfiguring Robots

Developer: Franz Nigl, Shuguang Li, Jeremy E. Blum, and Hod Lipson

The structure reconfiguring robot can act as a single entity or as a swarm and move along truss structures to assemble, disassemble, and position and reposition individual truss elements (Figure 2.185). The system can be used to build up any desired building structure from truss elements. The robot consists of two identical base bodies connected by an activated joint in the middle. Furthermore, each of the base bodies contains a mechanism for translational movement along the trusses and rotational movement around the trusses. Equipped with these mechanisms the robot can move/rotate along trusses (mechanism for translational/rotational movement; Figure 2.186), and it can move across corners and assemble/disassemble truss elements (Figure 2.187). To assemble/disassemble truss elements the two base bodies of the robot form a perpendicular structure through the activated joint in the middle. One base body then holds the robot on the structure in the desire position, and the translational/rotational mechanism of the other based body screws the truss element

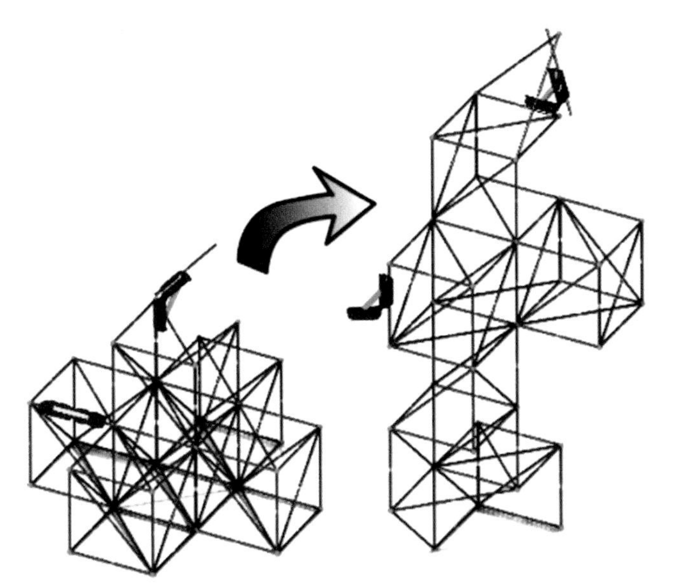

Figure 2.185. Schematic representation of several robots composing and decomposing truss structures. (Image: © [2009] IEEE. Reprinted, with permission, from Yun et al. 2009.)

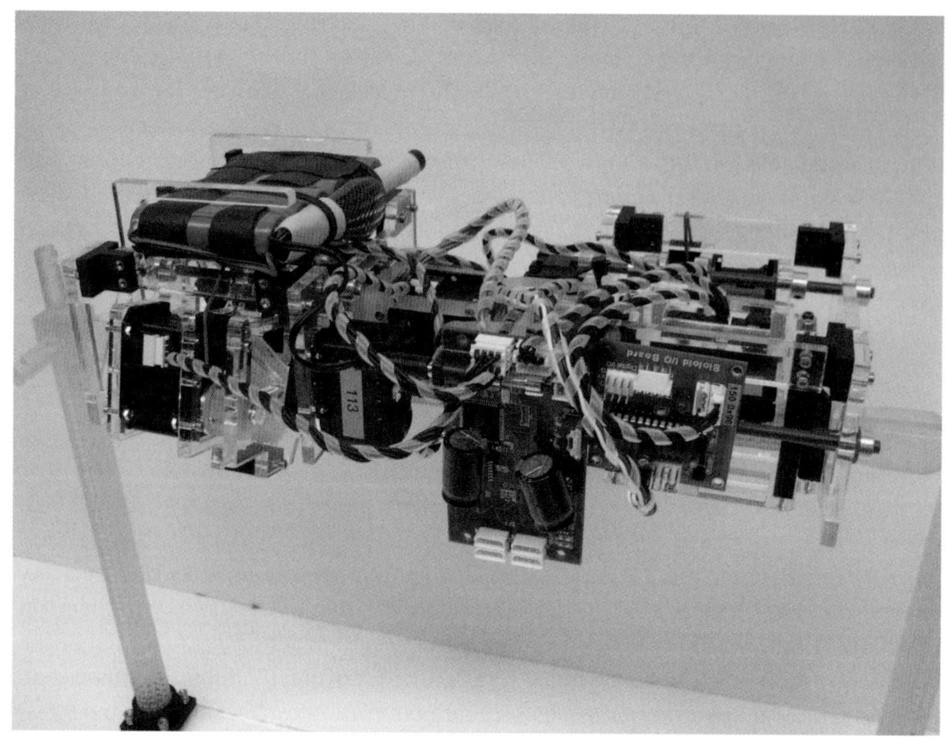

Figure 2.186. Prototyped scale model moving along a truss element. (Image: Jeremy E. Blum.)

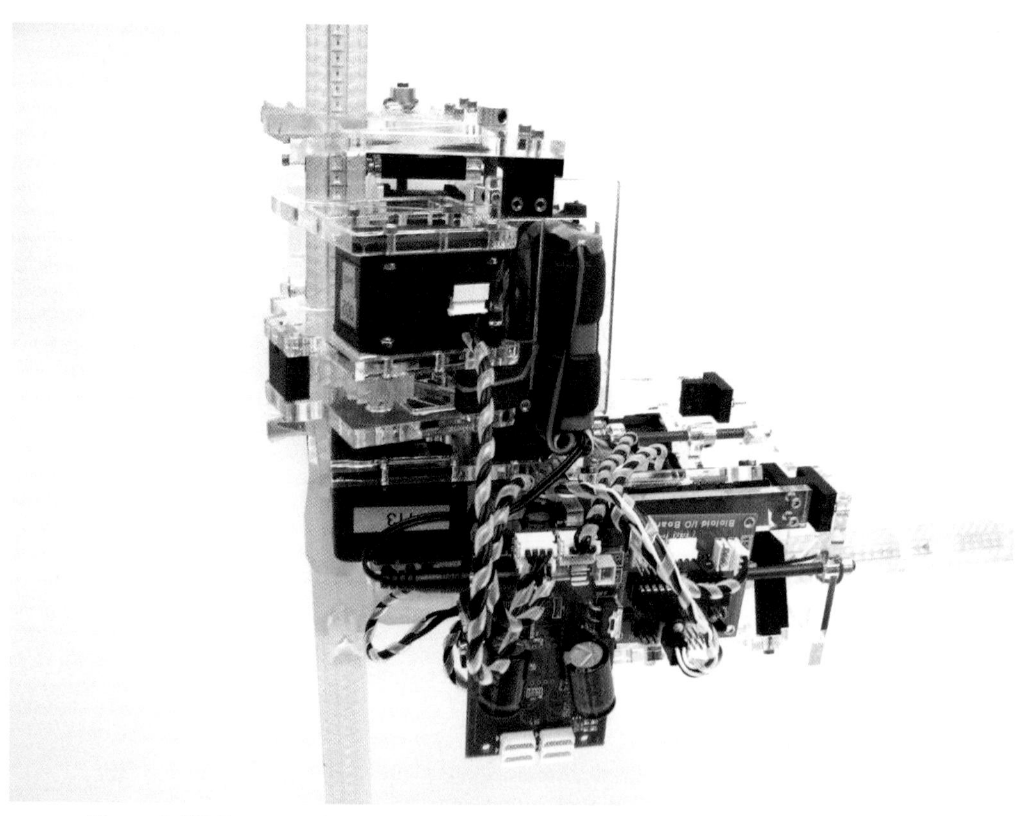

Figure 2.187. Prototyped scale model in rectangular mode assembling a truss element. (Image: Jeremy E. Blum.)

to be assembled into the joining system. The truss element kit is considered as an integral part of the system and the truss elements are designed to facilitate the operation of the robot. The truss elements are 3D printed and contain a structure that allows the gear wheels of the translation/rotation mechanism a safe grip in all directions. Furthermore, the connectors are designed to simplify the joining operation for the robot. Both robot and truss structure were prototyped as a scale model and tested in the laboratory to assemble simple structures.

For further information, see Nigl et al. (2013), Hjelle and Lipson (2009), Lobo et al. (2009), and Yun et al (2009).

2.14 Robots for Positioning of Components (Crane End-Effectors)

The transportation, elevation, balancing, and placement of materials in a conventional manner using a crane is not efficient and often leads to safety issues or material damages. Cranes are used primarily for logistics purposes, that is, transporting and placing materials on the floor or in the area where they are required. However, for positioning and alignment operations, the components must often be picked up again and handled by another system and the flow of materials is thus interrupted. Robotic positioning aids and robotic crane end-effectors improve conventional systems and methods, and allow for precise pick-up and position/alignment operations. Systems in this category range from relatively simple robotic end-effectors that allow only failsafe transport and the tele-controlled releasing of components to more complex systems that allow for rough positioning of a column or beam to highly complex multi-DOF end-effectors that allow accurate positioning and orientation of components. Some of the prototypes of the featured systems were in trials and demonstration also equipped with small, controllable turbines and gyroscopes to experiment with even more advanced features for position/orientation control. All the featured systems might serve as end-effectors for conventional cranes as well as for more advanced, automated logistics or carne solutions.

Robots for Positioning of Components (Crane End-effectors)		
87	Robotic crane end-effector Mighty Jack	Shimizu
88	Construction robot auto-shackle	Kwangwoon University & Samsung.
89	Robotic crane end-effector Robo-Crane	NIST
90	Robotic crane end-effector Mighty Shackle Ace	Shimizu
91	Autoclaw	Obayashi
92	Autoclamp	Obayashi
93	Robotic end-effector	Obayashi

System 87: Robotic Crane End-Effector Mighty Jack

Developer: Shimizu Corporation

The robotic crane end-effector Mighty Jack by Shimizu is a steel beam positioning manipulator used for the placement of medium-sized steel beams (Figure 2.188). It has dimensions of L 6.70 – 7.80 m × W 1.00 m × H 1.40 m, a weight of

Figure 2.188. Mighty Jack system in action. (Image: Shimizu Corporation.)

1800 kilograms, and a hanging load capacity of 1500 kilograms. The end-effector can be attached to conventional as well as advanced, automated cranes, lifting mechanisms, or overhead manipulators (OMs), and allows improving the cycle time and safety of the assembly procedure. The end-effector (activated by hydraulic activators) contains two grippers at its ends that allow it to be fixed between two columns, a lifting mechanism to handle (including lift and lowering) of the steel beams to be placed, and a telescopic mechanism that allows it to adjust to a variety of distances between columns. The system is lifted into place (already holding the beam to be placed), for example, by a tower crane. The two grippers at the ends of the end-effector allow the system then to attach to the columns of the already built steel structure segments, use them as a reference framework for the beam positioning, and adjust its position. Following this, the steel beam is lowered by the end-effector in a controlled manner to the desired position between the two columns and the beam component released automatically once demanded. The steel beam is fixed to the surrounding steel structure by conventional labour-based methods. The control approach is a hybrid between fixed sequence control and wireless tele-operation.

System 88: Construction Robot Auto-Shackle

Developer: Samsung and Kwangwoon University (Department of Electrical Engineering)

The construction robot auto-shackle is used for the placement of large steel columns and beams (Figure 2.189). The traditional method of placing these components requires labourers to connect and disconnect the members from the crane shackle manually. This method is time consuming and dangerous, especially at great heights. The auto-shackle system allows this process to become automated and controlled remotely. The system consists of a main unit (with an integrated communication module, a control unit, and a rechargeable battery for power supply) and two clamps that through a controllable pin allow automatic release of steel columns once brought into the desired position (Figure 2.190). It is controlled remotely via wireless communication through a control panel. The system is equipped with a limit switch that detects the locking state of the shackle pin. The improvement of the system in terms of miniaturization and clamping mechanism towards an advanced auto-shackle was demonstrated by Yi et al. (2004).

For further information, see Yi et al. (2004).

Figure 2.189. Auto-shackle ystem in action. (Image: Samsung and Kwangwoon University.)

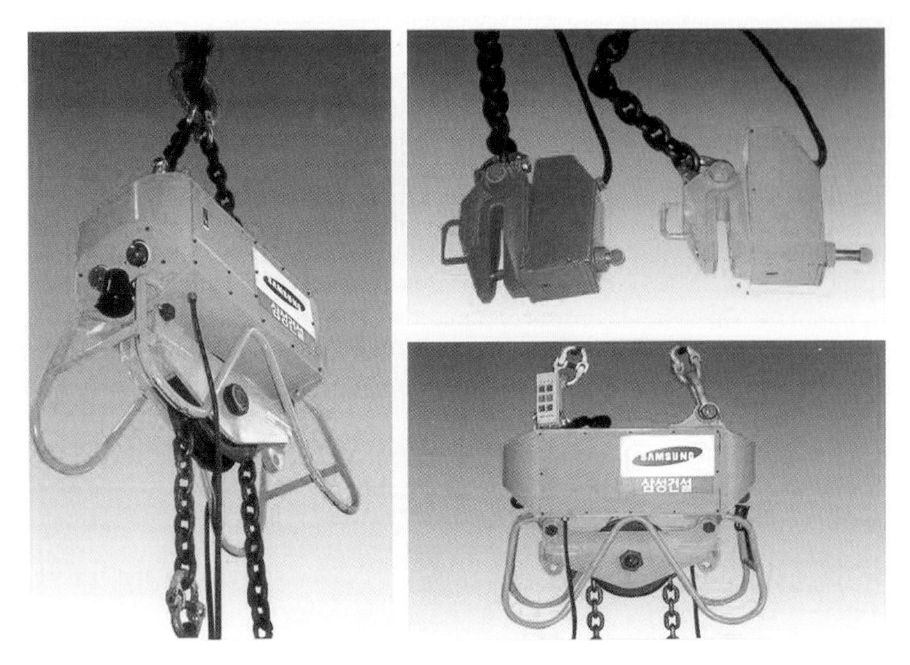

Figure 2.190. Detailed view of auto-shackle system. (Image: Samsung and Kwangwoon University.)

System 89: NIST RoboCrane

Developer: National Institute of Standards and Technology (NIST)

The NIST robot crane end-effector adds robotic features (in the form of the ability to precisely position and orient steel components) to conventional cranes.

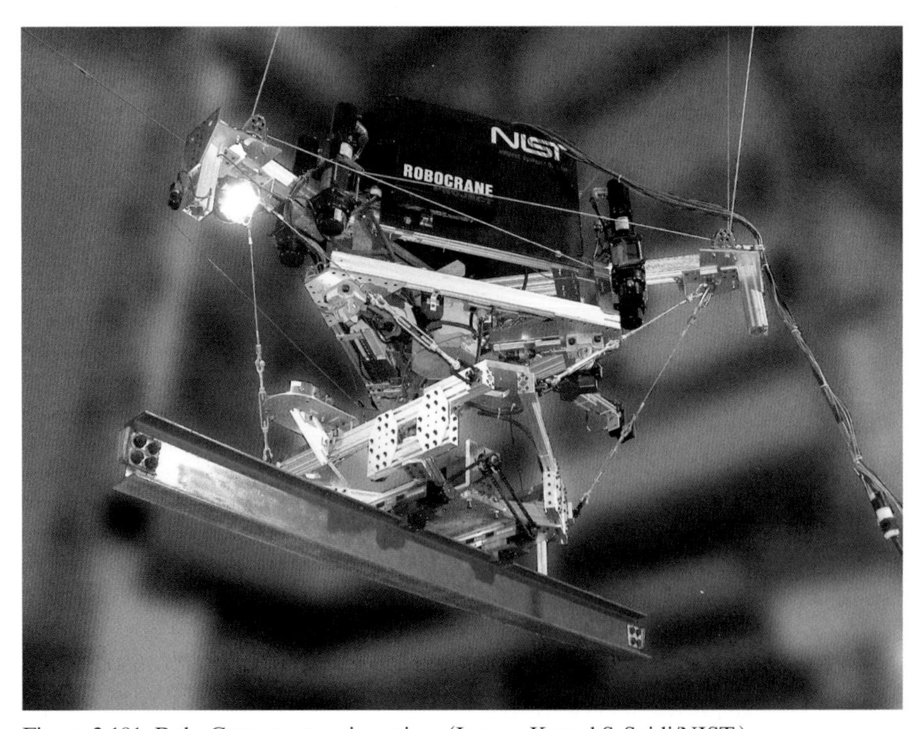

Figure 2.191. RoboCrane system in action. (Image: Kamel S. Saidi/NIST.)

The end-effector consists of a base unit (that integrates battery pack, control unit, etc.), a 6-DOF parallel kinematic manipulator (inverted Stewart-Gough Platform), as well as a robotic gripper (Figure 2.191). The end-effector allows one to both stabilize the load (by a real-time control system) as well as to position and orient it exactly in a partly automated, tele-operated manner. A variety of control mechanisms were developed through the different versions of the system ranging from a 6-DOF joystick-based operation to an automated operation based on higher level command input were tested and implemented. One of the versions developed allowed, for example, an autonomous assembly constructed using a laser-based on-site measuring system and assembly procedures derived from the CAD data.

For further information, see Saidi et al. (2006).

System 90: Robotic Crane End-Effector Mighty Shackle Ace

Developer: Shimizu Corporation

The Mighty Shackle Ace is a robotic end-effector that can be attached to conventional cranes to assist with the installation of steel components on the construction site (Figure 2.192). The system consists of a main unit (that integrates control and

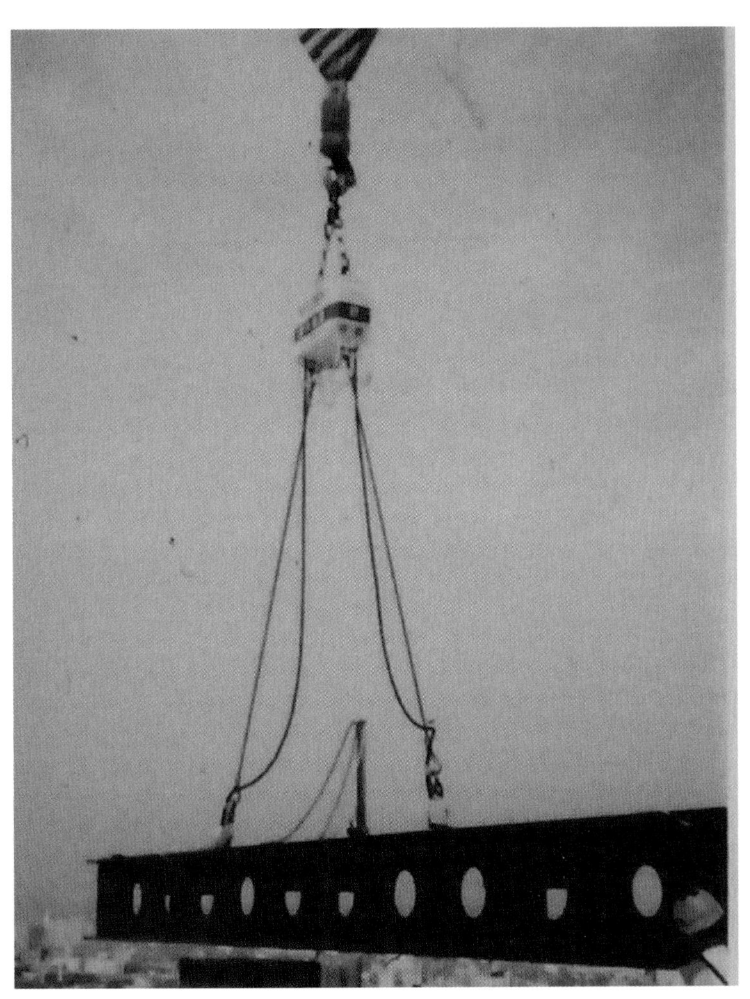

Figure 2.192. Robotic crane end-effector Mighty Shackle Ace. (Image: Shimizu Corporation.)

power units, stroboscopic lights, and an antenna system), a lifting wire attached to the top of the main unit to allow installation of the system to a crane hook, a leg system attached to the bottom of the system (that allows one to place and store the system on the ground), and two lifting wires each equipped with a controllable clamp (each clamp is additionally connected to the main unit through a cable that allows for signal transmission for clamp control). The system weighs about 250 kilograms, is able to handle relatively large components weighing up to 15 tons, and is wirelessly remote controlled.

For further information, see Ueno et al. (1988).

System 91: Auto-Claw

Developer: Obayashi Corporation

Obayashi's Auto-Claw (Figure 2.193) is a robotic end-effector that can be attached to standard and nonrobotic cranes, thus adding some basic robotic features to their performance. The Auto-Claw allows a crane operator to pick-up, install, and release, in particular, mid-sized steel beam segments in a controlled manner. It simplifies the positioning/orientation and fine-positioning of steel beam segments. The Auto-Claw speeds up the construction process and improves safety on the construction site as it eliminates the need for a worker to access the column and beam segments during the positioning process. The carrying capacity of the system is approximately 2 tons. A fail-safe system ensures that the lifted segments are not accidentally released in case of a sudden loss of power.

Figure 2.193. Auto-Claw. (Image: Obayashi Corporation.)

Figure 2.194. Auto-Clamp. (Image: Obayashi Corporation.)

System 92: Auto-Clamp

Developer: Obayashi Corporation

Obayashi's Auto-Clamp (Figure 2.194) is based on the same principles and technologies as Obayashi's Auto-Claw. However, in contrast to the Auto-Claw, this crane end-effector is designed for the installation of heavier elements such as large steel columns. The Auto-Clamp is a robotic end-effector that can be attached to standard and nonrobotic cranes, thus adding some basic robotic features to their performance. The Auto-Clamp allows a crane operator to pick up and release, in particular, large column segments and it simplifies the positioning of those segments. The Auto-Clamp speeds up the construction process and improves safety on the construction site as it eliminates the need for a worker to conduct work at the top end of the column during the positioning process. The carrying capacity of the system is approximately 14 tons. A fail-safe system ensures that the carried segments are not accidentally released in case of a sudden loss of power.

Figure 2.195. Robotic end-effector for Big Canopy.

System 93: Robotic End-Effector for Big Canopy

Developer: Obayashi Corporation

This robotic end-effector (Figure 2.195), developed by Obayashi for its automated/robotic on-site factory, Big Canopy, is conceptually and in terms of embedded technology similar to the Auto-Claw and the Auto-Clamp. In contrast to the Auto-Claw and the Auto-Clamp, this robotic end-effector is designed, in particular, for the handling of various types of prefabricated concrete elements. Similar to the other systems mentioned, this system consists of robotic end-effectors that can be – aside from its utilization within the Big Canopy – attached to standard and nonrobotic cranes, thus adding some basic robotic features to their performance. The system allows a crane operator to pick-up, position, orient, and release concrete elements.

2.15 Steel Welding Robots

The construction of larger buildings that use steel as the primary material for the bearing structure involves, in particular, a large amount of welding lines. If the design of the posts and beams allows for a reduction in the variety of welding lines, welding becomes a highly repetitive operation suitable for automation. Furthermore, conventional labour-based welding – especially when the provided equipment is of low quality – can have severe adverse effects on workers' health. Welding is a potentially hazardous job and precautions are required to avoid burns, vision damage, inhalation of poisonous gases and fumes, and exposure to intense ultraviolet radiation. In addition, the job must be done with high accuracy, often in uncomfortable working positions. Automated welding is able to better control and guarantee the quality of the connection between welded parts. Simultaneous automated welding on a beam (e.g., from two or more different but coordinated positions) is even able to ensure that the steel component does not become distorted and thus guarantees a high degree of accuracy. Systems in this category include small-scale systems that can temporarily be attached through a ring or template along which they operate to a beam or column, mobile platforms (equipped with welding systems) that can be driven to the column or beam joint to be welded, and larger ceiling suspended systems. Both in Japan and Korea the fact that several companies are active both in shipbuilding and construction has fostered the transfer of robotic steel welding approaches from ship building to construction.

Steel Welding Robots		
94	Steel beam welding robot	Obayashi
95	Steel column welding robot	Shimizu
96	Steel column welding robot WELMA	Fujita
97	Welding machine for steel girders	Kawasaki
98	Steel column welding robot	Obayashi
99	Steel beam welding robot	Fujita
100	Small sized column welding robot AUWEL 3	Kawada
101	Robotic welding system for the joining of large steel assemblies	Kawasaki Heavy Industries, Ltd., Robot Division

System 94: Steel Beam Welding Robot

Developer: Obayashi Corporation

Obayashi originally developed this system as a subsystem of its automated building construction system (ABCS; see also **Volume 4**) for automating the welding of beams to columns. Within the ABCS, the OM places the system on the beam adjacent to the column. However, it can also be used on more conventional construction sites, as positioning and repositioning of the system can be done by a conventional crane as well. The robot system consists of a frame, two manipulators mounted on top of this frame, an end-effector for welding, and a power supply system (Figures 2.196 and 2.197). On the one hand, the frame ensures that the system can be placed and fixed on a beam that supports it during the welding process. On the other hand, the frame can also lift the manipulators using up-and-down translational movement for approximate positioning. The manipulators add additional flexibility to the system and allow it to handle various shapes and sizes of both beams and joints.

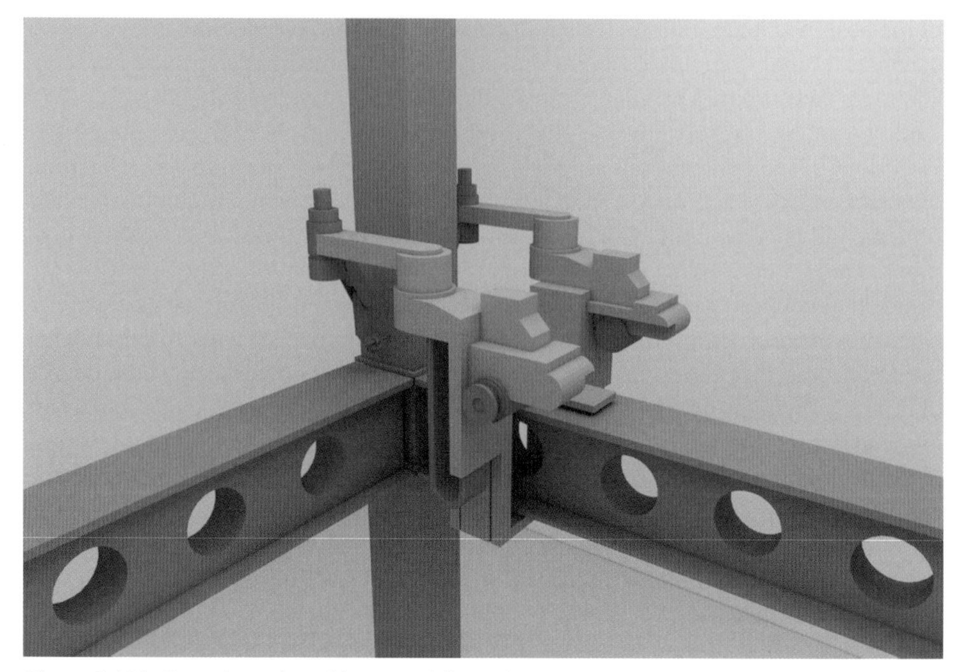

Figure 2.196. Overview of steel beam welding robot system.

System 95: Steel Column Welding Robot

Developer: Shimizu Corporation

The system is used for the automated welding of horizontal joints of steel columns. The system reduces manual labour as well as welding deficiencies due to human error. The system consists of a mobile carriage that carries a supply system/power source (and that can through a set of cables supply several welding robots

Figure 2.197. Steel beam welding robot system in action. (Image: Obayashi Corporation.)

Figure 2.198. Column welding robot. (Image: Shimizu.)

attached to different columns), a guiderail for the robot that is temporarily fixed to the column, and the welding robot itself (Figure 2.198). The robot is powered by a DC servomotor, travels around a column via the temporary guiderails, and is equipped with a welding end-effector. The robot is also equipped with an advanced welding control system able to detect the shape between the columns using laser sensing. Based on the scan of the joint the robot does, it can select the appropriate profile from its database and proceed to weld the columns together. The system is equipped with internal monitoring capability and can, for example, determine a shortage of welding wire or malfunctions during welding. The system requires a manual mounting of both the guiderail and the robot to each specific column. Once initiated, as a result of the guiding rail approach, the robot operation is simple and robust and therefore a single welder can manage several welding robots attached to different columns simultaneously.

For further information, see Council for Construction Robot Research (1999, pp. 196–197).

System 96: Steel Column Welding Robot WELMA

Developer: Fujita Corporation

This system was developed for the welding of the joints of a building's structural columns. Columns with a size of 450–800 millimetres and a thickness of 22–80 millimetres can be welded by the robot with a speed of 300–600 mm/minute. The basic

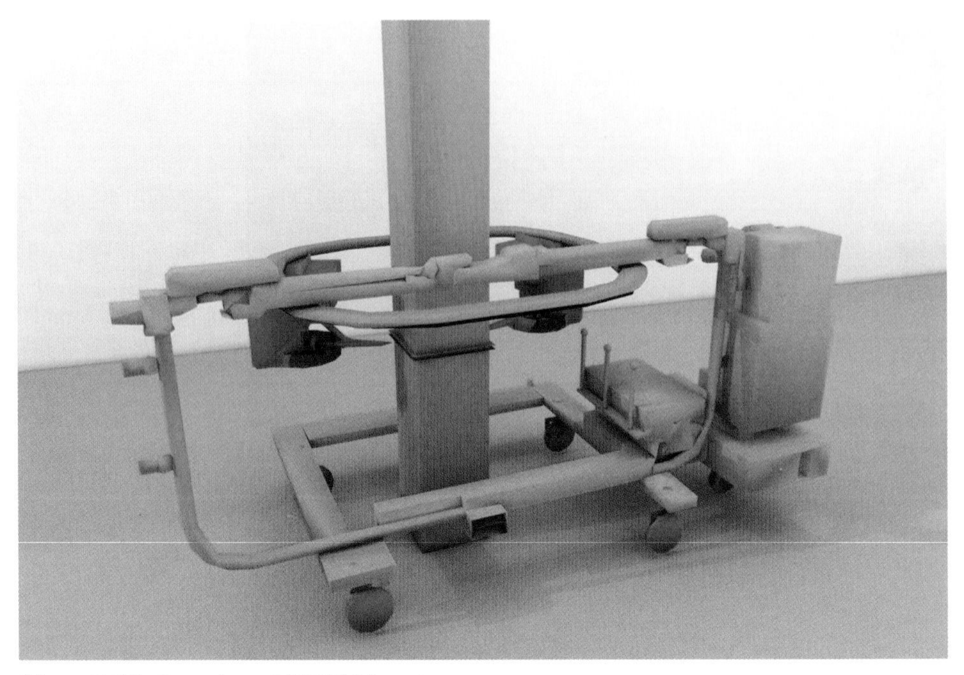

Figure 2.199. Overview of WELMA system.

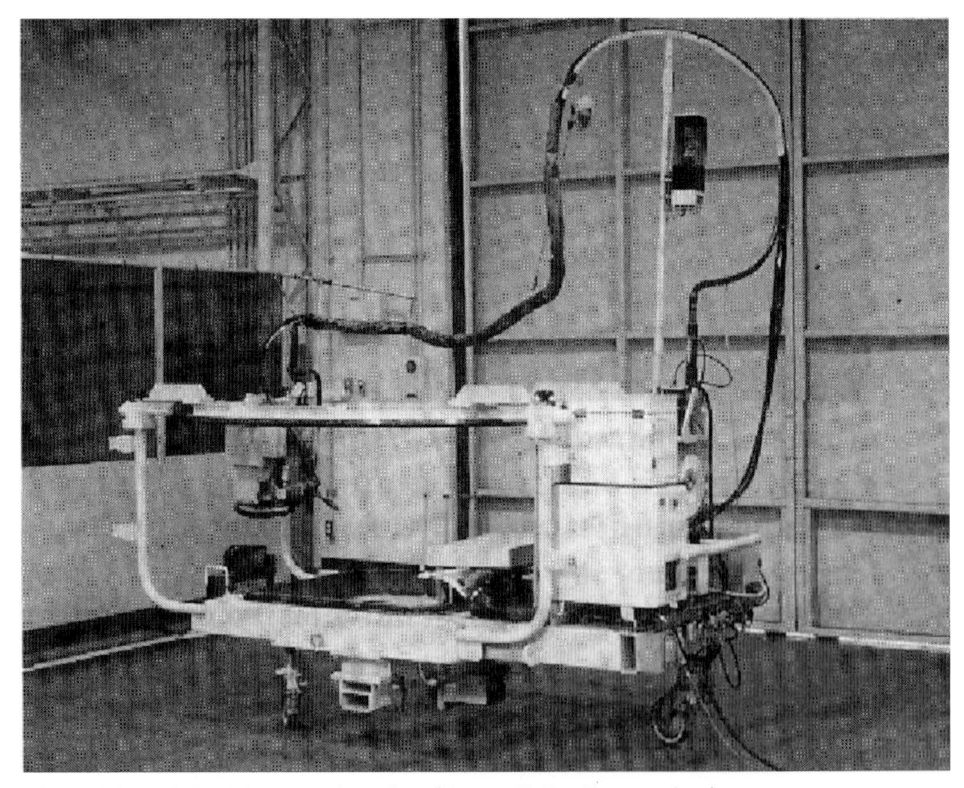

Figure 2.200. WELMA system in action. (Image: Fujita Corporation.)

element of the robot is a manually movable mobile frame that can be opened and closed and thus allows the robot to grasp the columns to be welded. This frame carries guiding rails on which the two manipulators carrying the welding end-effector travel in a coordinated manner around the column as well as an on-board power supply system (Figures 2.199 and 2.200). After the installation of the frame at the required column location, the robot performs the multilayer welding task by automatically detecting the welding grooves with its sensors. The ability to weld the columns through the system simultaneously from various sides reduces the possibility of distortions of the columns through the welding process.

For further information, see Council for Construction Robot Research (1999, pp. 172–173).

System 97: Welding Robot for Steel Girders

Developer: Kawasaki

The welding robot for steel girders from Kawasaki (work area: W 6.00 m × L 18.00 m) is used to weld the stiffeners of the steel girders. The system consists of two large robotic welding robots suspended from a gantry crane structure able to travel along a two-rail track (Figure 2.201). The gantry crane allows for the suspended robots to travel in both the *x*- and *y*-directions. Two Kawasaki Js-6 welding robots, equipped with welding end-effectors, provide additional degrees of freedom and carry out the welding. The system is equipped with a number of sensors. A charge-couple device camera is used for edge detection while a laser sensor allows for active position detection. A light sensor allows for the detection of start and end positions. An arc sensor that tracks the seam of the weld is used as the basis for quality control. A safety system stops the system when the presence of a human being is

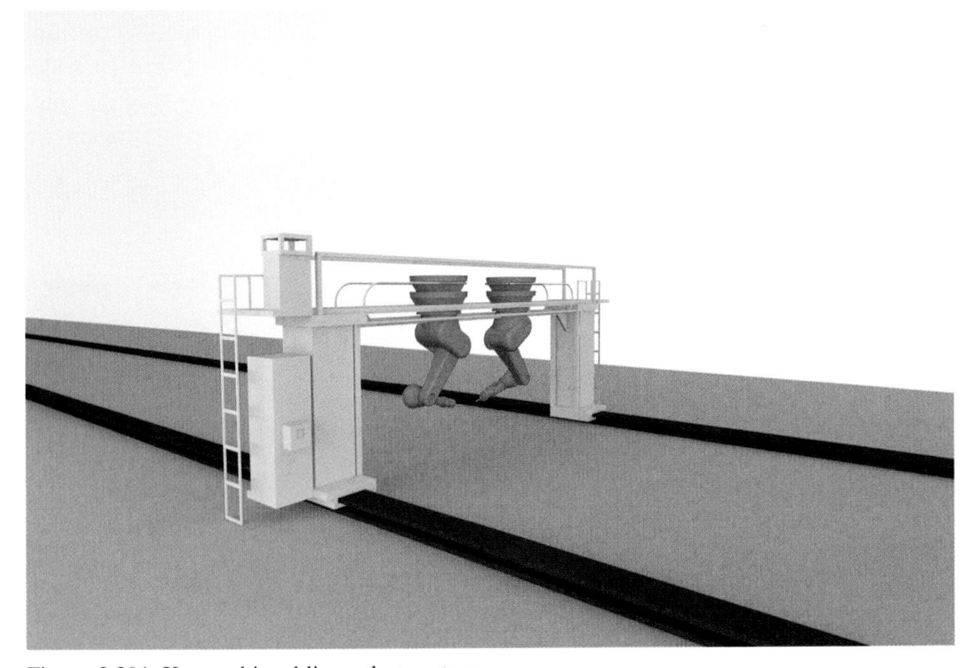

Figure 2.201. Kawasaki welding robot system.

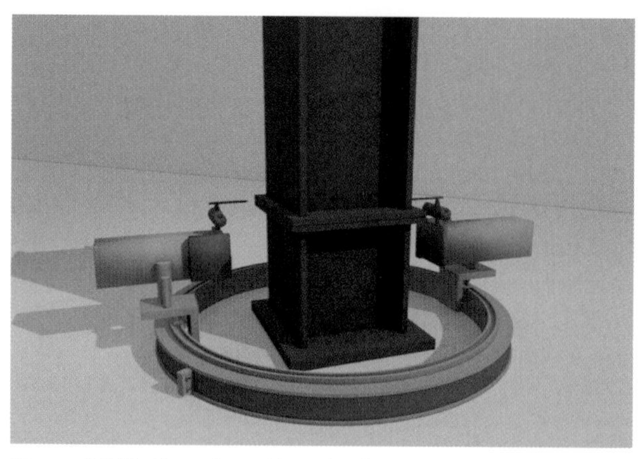

Figure 2.202. Overview of steel column welding robot system.

detected in the work area. The girders to be welded must fall within the following ranges: length: 4–14 metres, width: 1.5–3.5 metres, height: 100–300 millimetres, and thickness: 9–22 millimetres.

For further information, see Council for Construction Robot Research (1999, pp. 176–177).

System 98: Obayashi Steel Column Welding Robot

Developer: Obayashi Corporation

Obayashi developed this system as a subsystem of its automated building construction system (ABCS; see also **Volume 4**) for automating the welding of joints between columns. The system can be used within the ABCS as well as on more conventional construction sites. The system is driven to the column joint to be welded. It welds the column simultaneously from opposite directions to minimize distortion of the column. The system consists of a mobile unit, a ring-like frame connected to the mobile unit, two manipulators that can move along the ring-like frame, and a supply unit (Figures 2.202 and 2.203). The ring-like frame can be

Figure 2.203. Steel column welding robot system in action. (Image: Obayashi Corporation.)

opened and closed, thus providing a continuous path around the column for positioning the manipulators. The manipulators are flexible and allow a variety of column and joint types to be welded. Once set up, the welding process is conducted fully automatically.

System 99: Steel Beam Welding Robot

Developer: Fujita Corporation

Fujita's mobile welding robot can be used within its automated/robotic on-site factory as well as on conventional construction sites. The system was designed primarily for welding of beam segments to individual columns. The robot system consists of a mobile platform, a 5-DOF manipulator, a flexible working platform that can be temporarily positioned near the welding location, and a welding end-effector (Figure 2.204). The manipulator is used to position both the working platform and

Figure 2.204. Steel beam welding robot. (Image: Fujita Corporation.)

end-effector. Human workers can access the location where welding is conducted via the working platform. This allows them to (in the case of automatic operation) oversee the welding process or (in the case of manual operation) guide the welding end-effector. In the case of manual operation, the human worker can conduct complex welding operations as the manipulator conveniently holds and repositions the end-effector, thus allowing the human worker to concentrate on the actual welding process.

System 100: Small-Sized Column Welding Robot AUWEL 3

Developer: Kawada Corporation

AUWEL 2 was an experimental system that was developed and tested under laboratory conditions and was developed further towards the AUWEL 3 (Figure 2.205). The aim of the system was to automate the welding of large column segments to each other. The system consists of a fixture-like frame attached to the column equipped with guiding profiles along which the manipulators can travel to access the areas to be welded around the column. A welding end-effector is attached to those manipulators. The fixture-like frame ensures a fast installation and deinstallation of the system to columns on the site. The system with this fixture structures the environment for the robotic manipulators by providing a reference frame. AUWEL robots can weld the columns from two sides simultaneously and in a coordinated manner, thus minimizing the risk of major distortions and misalignments caused by the welding process. It is intended that multiple AUWELs are deployed on the construction site in parallel to significantly speed up the erection of steel frame structures.

Figure 2.205. Small-sized column welding robot AUWEL 3. (Image: Kawada Corporation.)

System 101: Robotic Welding System for the Joining of Large Steel Assemblies

Developer: Kawasaki Heavy Industries, Ltd., Robot Division

Kawasaki developed this robot system for the joining of large steel assemblies. Currently the application of the system targets the construction of large ships. However, as mentioned in **Volume 1**, many similarities exist between the construction of ships and construction of buildings. With some modifications it is imaginable that such systems could be used in on-site building construction and apply the systems or individual subsystems in prefabrication (off-site) or on site in automated/robotic on-site factories or as single, distributed systems. Indeed, in particular in the context of automating welding tasks, automated building construction has previously often built on approaches coming from the ship building industry. The robotic welding system consists of several elements. Off-the-shelf Kawasaki robots are equipped with welding end-effectors and are placed in a frame (Figure 2.206) that can be lowered by overhead cranes equipped with telescopic arms (Figure 2.207). The frame contains both sensors and activators that allow besides automated welding an automated fine positioning of the frames within the steel structure where the welding operations have to be performed.

2.16 Facade Installation Robots

Facade installation operations include the positioning and adjustment of windows, complete facade elements, or exterior walls of a building. Facade elements, in modern

Figure 2.206. SRL variant 1 – strong limbs worn by an assembly worker provide assistance from the hip area for handling and installing large and heavy panels. (Image: Kawasaki Heavy Industries, Ltd.; Kawasaki Robotics Division.)

Figure 2.207. SRL variant 2 – highly dexterous limbs provide assistance with the overhead installation of interior elements. (Image: Kawasaki Heavy Industries, Ltd.; Kawasaki Robotics Division.)

architecture and especially in high-rise construction, are decoupled from the bearing concrete or steel main structure and can thus be considered as a type of "infill" or "platform" system (for further explanation of infill and platform approaches, see also **Volume 1**). Facade installation operations are complex operations that involve the accurate positioning of heavy parts or elements at locations that are difficult to access (e.g., high altitudes without scaffolding). This involves the risk of injury (and thus requires extensive safety measures) and of damaging expensive elements. Furthermore, the positioning and alignment of prefabricated facade elements requires precision with low tolerances for error. Since the 1980s, the growing trend of designing large buildings as monolithic structures repeating similar facade elements has provided a major incentive for investment into the development of automated or robotic systems. Up to the present day, facade installation systems have been a hot topic in R&D departments, especially firms in Asia where high-rise buildings are becoming more and more prevalent, even in residential construction. Systems in this category include mobile robots that can be used on the individual floors to install facade components and elements to the building from the floor level, highly mobile spider-like robotic cranes for the installation of elements, gondola type systems that operate along the roof of the building from where thy suspend and install the panels, as well as robots used to install roof covers in an automated manner. On the very top end the systems of Fujita and Brunkenberg automate logistics as well as positioning and alignment operations.

Facade Installation Robots		
102	Robot for the installation of post and beam structures of facades	Kajima
103	Facade element installation robot	Kajima
104	Robot for positioning facade elements from upper levels on lower levels	Kajima
105	Facade concrete panel installation robot	Kajima
106	Automatic facade panel installation using a movable fixture (right positioning)	Fujita
107	Facade element installation robot	Hanyang University & Samsung
108	Roof cover installation robot	Kumagai Gumi
109–110	Various types of on-site material handling systems • Spider-like mini cranes • Glazing robots and end-effectors	GGR Group
111	Ceiling glass installation robot	Hanyang University & Samsung
112	"Shuttle system" for the installation of facade elements	Fujita
113	Brunkeberg® system	Brunkeberg Systems AB
114	Robotic installation systems for facade panel installation in building renovation	Technische Universität München

System 102: Robot for the Installation of Post and Beam Structures of Facades

Developer: Kajima Corporation

Kajima developed this robot to assist workers in the installation of heavy post and beam elements of facades from the individual floor levels without the need for scaffolding or cranes. The robot consists of a mobile, tracked platform, a manipulator, and an end-effector. The mobile tracked platform allows the robot to be moved by the workers on the site and in particular on the floor levels to the desired locations. The manipulator can be operated by a worker in a cooperative manner and is controlled through control buttons at the end of the manipulator near the end-effector. The end-effector provides a grasping mechanism to pick up the post and beam elements (Figure 2.208) and release them once positioned correctly (Figure 2.209).

System 103: Facade Element Installation Robot

Developer: Kajima Corporation

The installation of facade elements can be a hard and dangerous task for human workers. It requires dealing with heavy panels and installing them carefully in the correct location and position. It often requires three to four workers inside the building

Figure 2.208. Picking up of post and beam elements of the facade with the end-effector of the robot. (Image: Kajima.)

and a crane working outside of the building, carrying the panel to the desired floor and holding it until the workers finish the installation process. The facade element installation robot by Kajima makes this process much easier and cheaper. It reduces the number of workers needed inside the building to one or two and ends the need for cranes to position the load from outside the building. The robot has a compact design and is movable, allowing it to be easily transported to the site and to the desired floor using conventional on-site construction lifts. The robot consists of a mobile platform that can lift and lower a manipulator and an end-effector that can pick up, hold, and lower facade panels through a winch mechanism (Figure 2.210). The robot allows thus to lift the elements into the desired position and keep them precisely in that position until a worker can fix it in place (Figure 2.211).

System 104: Robot for Positioning Facade Elements from Upper Levels on Lower Levels

Developer: Kajima Corporation

Although most facade elements can be installed from inside the building, when the facade elements are larger in size, there is not enough space and they must be installed from the outside of the building. The Kajima robot for positioning facade elements is used precisely for this job without the need for a crane. This is beneficial as cranes sometimes encounter problems when the construction site is surrounded by narrow streets. The robot uses an end-effector with a winch mechanism, and operates

Figure 2.209. Installation of post and beam elements of the facade to the buildings in a cooperative human–robot manner. (Image: Kajima.)

Figure 2.210. Facade element installation robot system operated on the finished floor. (Image: Kajima Corporation.)

Figure 2.211. Installation of facade panel. (Image: Kajima Corporation.)

Figure 2.212. Robot for positioning facade elements from upper levels on lower levels.

from the floor above the floor where the panels are to be installed (Figure 2.212). The robot is manually controlled by one skilled worker who can position the panel from outside of the building and hold it in place until it is fixed by another worker on the level below. The robot has a relatively heavy body, which allows it to deal with relatively large and heavy facade elements with high precision and ease.

System 105: Facade Concrete Panel Installation Robot

Developer: Kajima Corporation

Based on the Kajima LH series robots, Kajima's facade concrete panel installation robot system is equipped with a specialized end-effector for picking up, transporting, and positioning of comparably small facade panels. The robot system allows a single worker to conduct the installation process by manual remote-control of the system (Figures 2.213 and 2.214). The robot system consists of a mobile base frame, a 4-DOF manipulator, and an end-effector. The end-effector consists of a gripping unit equipped with vacuum suction devices and a winch-based lowering unit that allows the panel to be lowered to its intended location. The robot system allows for facade panels that are traditionally installed either by the tower crane (and thus blocking the major on-site logistics device) or by several construction workers in a risky operation to be installed by relatively convenient and safe method from within the individual floor levels.

System 106: Automatic Facade Panel Installation Using a Movable Fixture (Right Positioning)

Developer: Fujita Corporation

The system was developed for assisting and partly automating the installation of large concrete panels to multistory building structures. The system consists of

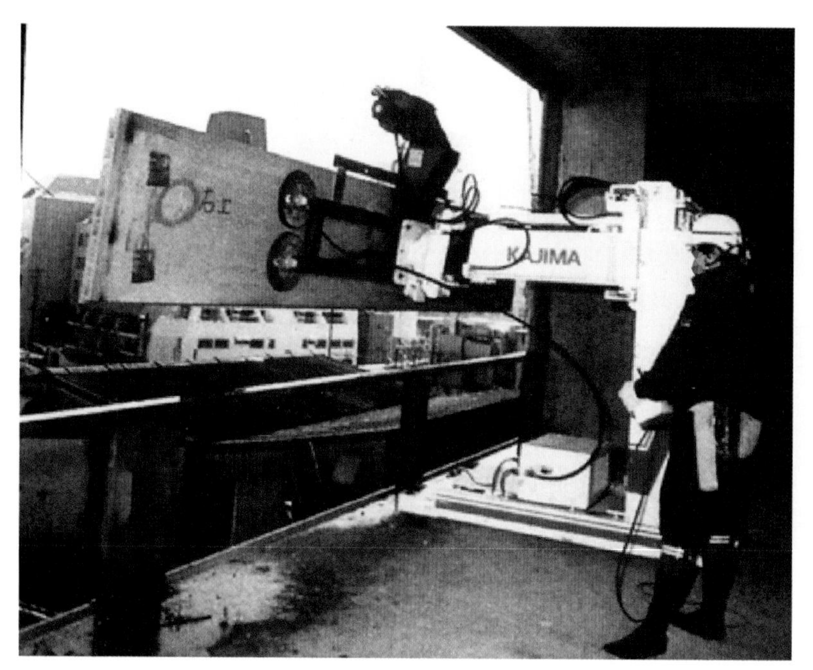

Figure 2.213. Facade concrete panel installation robot, Kajima Corporation.

guiding rails (Figure 2.216), an OM (with an end-effector- for handling the panels), and a movable, robotic template for panel fine positioning/adjustment (Figures 2.215-1 and 2.215-2). An OM travels along temporary rails installed on the building around the roof area. The OM elevates prefabricated panels from a predefined storage area located on the ground, lifts them vertically, delivers them horizontally to a position above the location where they shall be installed, and then lifts them down

Figure 2.214. Facade concrete panel installation robot, installing facade panel, Kajima Corporation.

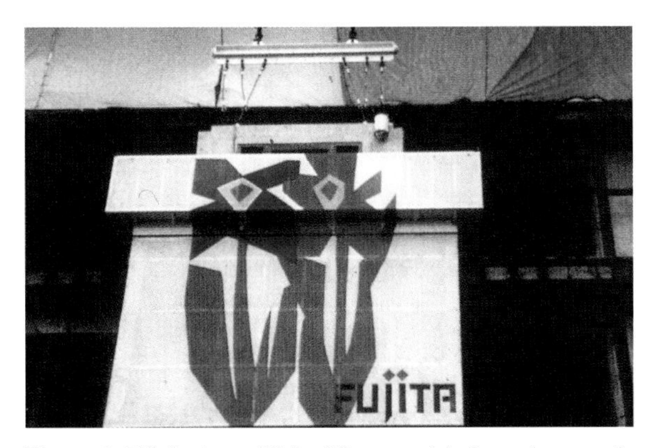

Figure 2.215-1. An OM lifts prefabricated panels down into a temporarily installed, movable robotic template that assists with positioning. (Image: Fujita Corporation.)

Figure 2.215-2. Once placed in the template by the OM, the robotic template uses a mechanism to push the panel towards the building and into its final position. (Image: Fujita Corporation.)

into the temporarily installed, movable template that assists with positioning of the panel. Once placed in the template by the OM, the template uses a mechanism to push the panel towards the building and into its final position. The system was used in an integrated manner in the context of automated/robotic on-site factories (see **Volume 4**) as well as a standalone system on more conventional sites.

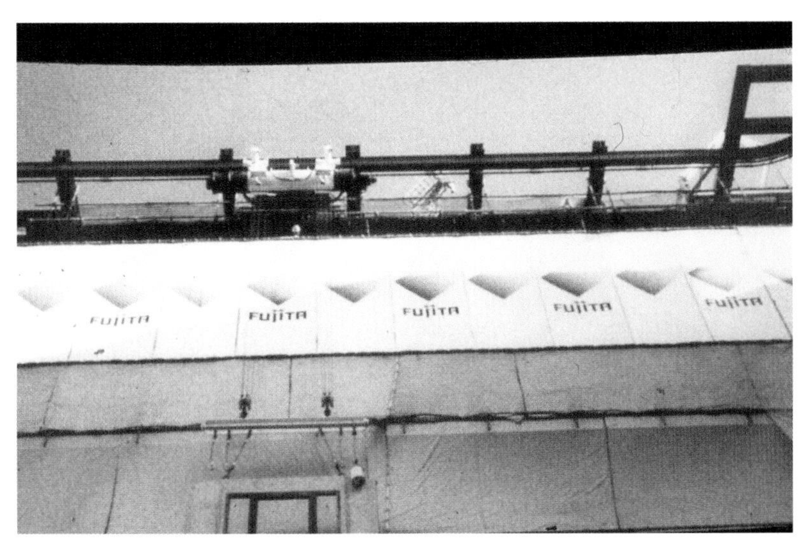

Figure 2.216. An OM delivers prefabricated panels to the robotic template. (Image: Fujita Corporation.)

Figure 2.217. Facade element installation robot system in action. (Image: Hanyang University.)

System 107: Facade Element Installation Robot

Developer: Hanyang University and Samsung

The system was developed for the installation of facade panels from the inside of the building (Figures 2.217 and 2.218) as opposed to the other systems that are designed for installation from the exterior. The system consists of a tracked mobile platform equipped with a manipulator that carries an end-effector equipped with controllable suction cups designed to handle facade panels. The manipulator is a robotic arm that contains five joints and four links, and is powered by means of a hydraulic system to lower, raise, extend, and retract the end-effector. The end-effector can accurately be controlled by the operator using force feedback and allows safe and accurate positioning and release of facade panels. In the future such systems shall be used extensively by Korean firms in high-rise construction, substituting conventional methods.

For further information, see Lee (2008).

Figure 2.218. End-effector for facade element installation robot. (Image: Hanyang University.)

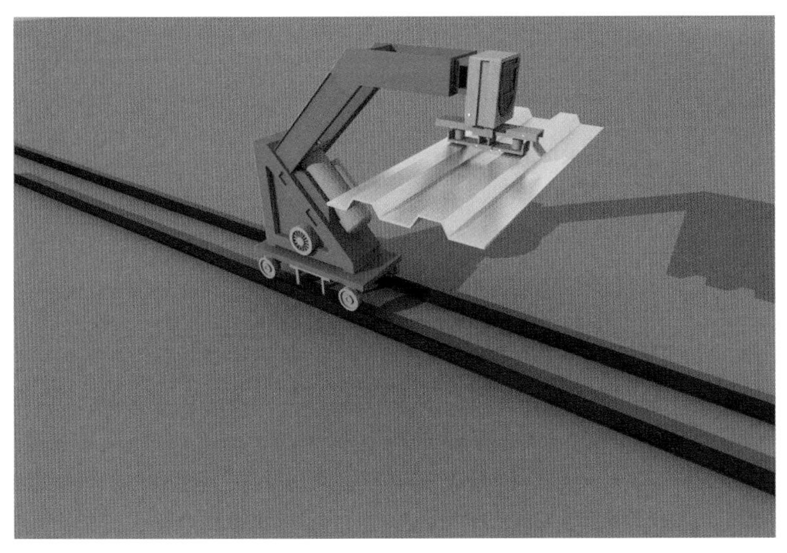

Figure 2.219. Overview of roof cover installation robot system.

System 108: Roof Cover Installation Robot

Developer: Kumagai Gumi

The system was developed to automate the installation of corrugated roof sheets on top of industrial buildings (Figures 2.219 and 2.120). The positioning of roof sheets

Figure 2.220. Roof cover installation robot system in action. (Image: Building Technology Development Department of Kumagai Gumi Corporation.)

is a highly repetitive task that also poses a safety risk for the workers involved and thus is predestined for being automated. The robot consist of a mobile platform, a simple manipulator (only 1 DOF: vertical movement of the last link along one axis), and an end-effector for handling the roof sheets. Owing to its small size and weight, the robot can move on rails installed on the trusses of the roof to transport and place the roof sheets just in time and just in sequence. Roof sheets are stored at dedicated locations from where the robot picks them up one by one, and after placing one sheet it travels back to pick up the next sheet and continues this task repetitively.

Systems 109 and 110: Various Types of On-Site Material Handling Systems

Developer: GGR Group

GGR Group is a UK-based company developing automated, cooperative equipment (e.g., pick and carry cranes, mini crawler cranes, mini spider cranes), robots (e.g., glazing robots), and end-effectors (e.g., vacuum lifters) for the handling of material on the construction site. For example, GGR offers a series of compact, spider-like mini cranes (Figure 2.221) covering a lifting capacity ranging from about 1 to 10 tons. The UNIC URW – 094 (width: 0.595 m; weight: 1000 kg, boom length: 5.50 m) allows the handling of loads up to 995 kilograms. This model of the series is so compact and

Figure 2.221. A UNIC mini spider crane with a glass vacuum lifter, installing glass panel onto an apex roof. (Image: GGR Group Ltd./UNIC Cranes Europe.)

Figure 2.222. A collage showing the OSCAR glazing robots. (Image: GGR Group Ltd.)

light that it allows to be operated both within the building and in the outdoor environment. A larger model of the series, the UNIC URW-1006 (lifting height: 30.7 m, working radius: 24.3 m) is capable of handling elements of up to 10 tons. The cranes can be remotely controlled through high-resolution colour liquid-crystal display screens. Furthermore, GGR offers a range of glazing robots (Figure 2.222) developed to assist in difficult glazing and curtain wall installation tasks through the inside of the building and via the individual floor levels in situations in which circumstances prevent the use of scaffolding or conventional cranes. For example, the OSCAR 600 features a glazing robot (lifting capacity: 600 kg/300 kg when the lifting arm is fully extended, reach of the telescopic manipulator: 1.445 m) with a chassis of a size of only 780 millimetres that is so compact that it can travel easily indoors and pass through narrow doors. The manipulator is controlled by a precision control system and a stabilizer is used to assist when handling heavier elements. GGR provides a variety of end-effectors (for tasks such as glazing panel, cladding, roofing, and ceiling installation) that can be attached to the GGR's mini cranes or robots as well as to conventional cranes of construction machines.

For further information, see GGR Group (2015).

Figure 2.223. Ceiling glass installation robot end-effector in action. (Image: Lee/ Hanyang University.)

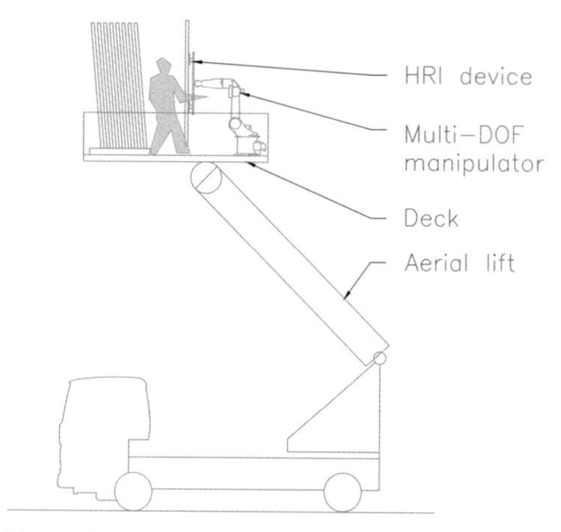

Figure 2.224. Overview of ceiling glass installation robot system.

System 111: Ceiling Glass Installation Robot

Developer: Hanyang University and Samsung

In general, building components and materials are quite bulky and can be difficult for manual handling on-site. Recently, ceiling glass panels are becoming a popular choice for architects when designing curtain wall elements and interior finish features. However, ceiling glass comes in different sizes and designs and can be difficult for manual handing, especially when carrying out overhead installation. The ceiling glass installation robot, jointly developed by Hanyang University and Samsung as a research project, aims to reduce labour requirements and improve productivity and safety. The system consists of three main parts: (1) a vehicle equipped with a lifting mechanism (max. payload = 2000 kg), (2) a work deck that can be positioned by this lifting mechanism, a multi-DOF industrial manipulator placed on the work deck, and (4) an end-effector equipped with a human robot interface (HRI) device that allows a worker to guide the element assisted by the industrial manipulator in place (Figures 2.223 and 2.224). The design of the system stresses the concept of human–robot cooperation in which a robot augments the power of human beings to flexibly install heavy panels. The vacuum suction device of the end-effector is designed to be compatible with a variety of ranges of glass sizes.

For further information, see Lee et al. (2007).

System 112: Shuttle System for the Installation of Facade Elements

Developer: Fujita Corporation

The shuttle system from Fujita is used for the easy installation of large exterior wall elements such as curtain wall panels (Figures 2.225 and 2.226). The system consists of a stage (end-effector) that surrounds the perimeter of the building, vertical guiderails for the stage to be lifted, and motor driven chain blocks that provides the ability to lift the stage. With this system, an entire floor of large curtain wall panels can be assembled to a facade segment in a dedicated area on the ground by human workers and then lifted up into place by the stage. This process is repeated facade segment by facade segment until the complete facade is installed. The system thus

Figure 2.225. Shuttle system in action. (Image: Fujita Corporation.)

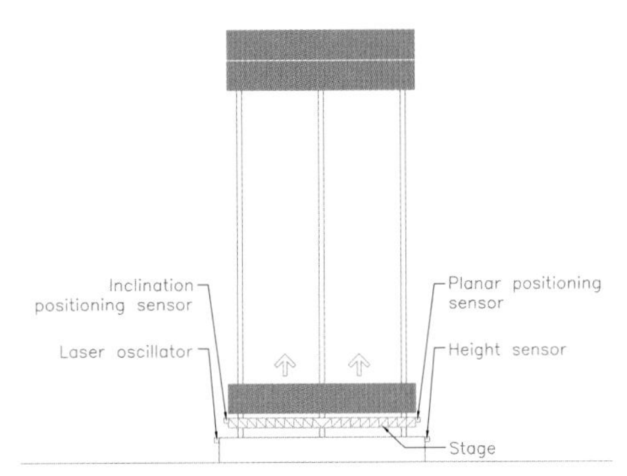

Figure 2.226. Overview of shuttle system.

eliminates a large amount of work that usually has to be done on the floor level and under inconvenient and dangerous work conditions and also eliminates the need for scaffoldings and cranes. The system is equipped with a laser distance sensor that measures the distance to the approaching stage five times per second to prevent collision and ensure a safe positioning and connection of the lifted facade segments to the segments above.

For further information, see Council for Construction Robot Research (1999, pp. 220–221).

System 113: Brunkeberg® System

Developer: Brunkeberg Systems AB

The Brunkeberg® system partly automates the logistics and installation of facade panels (Figures 2-227–2.235). Key parts of the system are the profiles attached to the main building structure for fixing the facade panels, the facade panels, and the logistics and installation mechanism. The profiles attached to the main building are designed to allow guidance of the mechanism for automated vertical delivery of the facade panels. However, the panels themselves also contain features that facilitate their handling by the logistics and installation mechanism. The panels are delivered by trucks to a logistics yard on the site where the logistics system picks them up and delivers them to a buffer storage. From there the panels are distributed hanging on rails by a horizontal delivery system (HDS) to the segment of the facade where they have to be installed. Guided by profiles attached to the main building structure, a vertical delivery system (VDS) takes them up from the HDS and delivers them vertically to an installation area where, assisted by a human worker, the panels are installed. Brunkeberg is backed by large companies such as Lindner Group

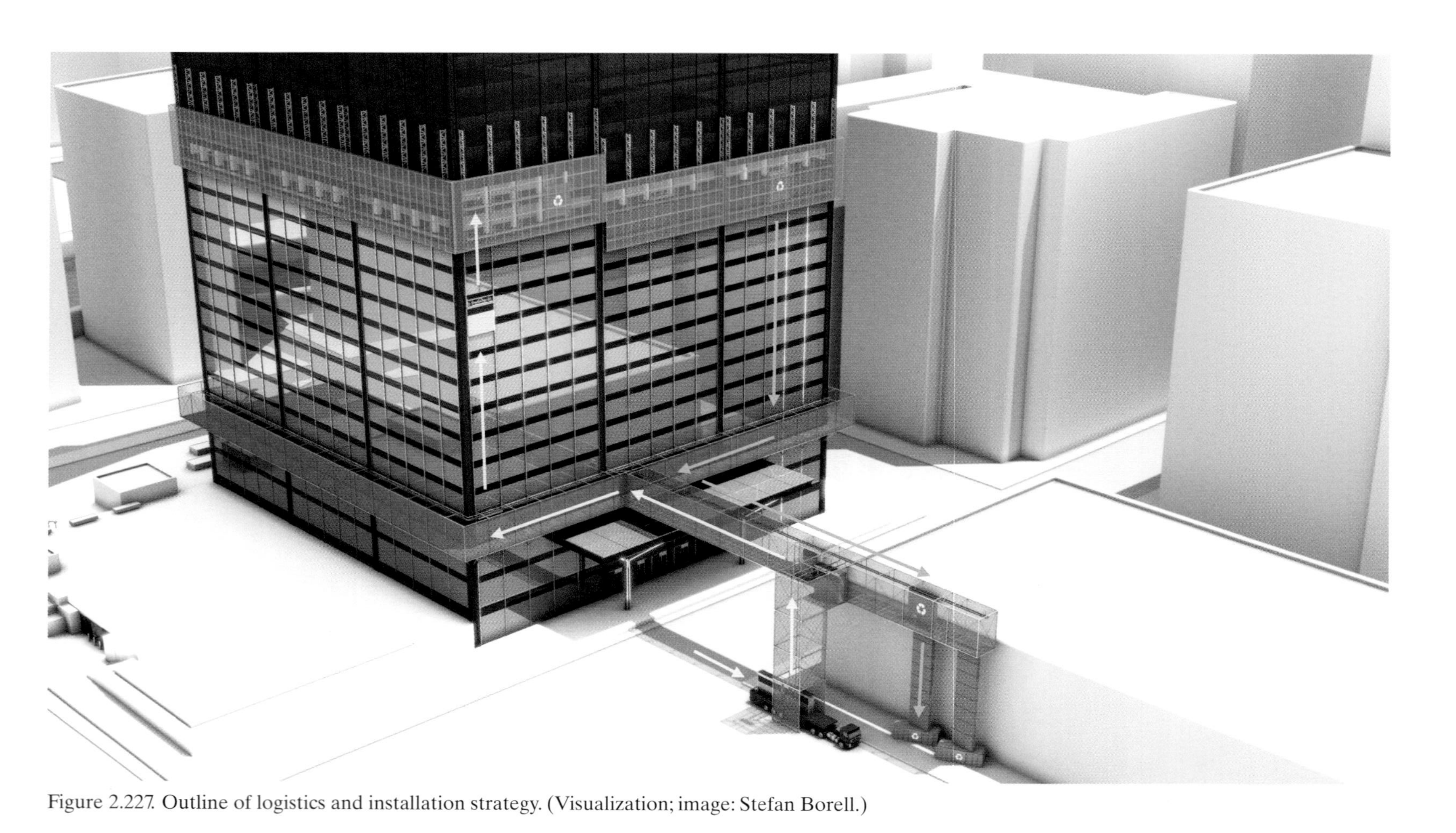

Figure 2.227. Outline of logistics and installation strategy. (Visualization; image: Stefan Borell.)

Figure 2.228. Material handling yard on the ground. (Visualization; image: Stefan Borell.)

Figure 2.229. Material handling yard on the ground. (Prototypic real-world application; image: Mattias Bergman.)

Figure 2.230. Horizontal panel distribution. (Visualization; image: Stefan Borell.)

Figure 2.231. Horizontal material distribution. (Prototypic real-world application; image: Mathias Farnebo.)

Figure 2.232. Vertical panel delivery and installation. (Visualization; image: Stefan Borell.)

Figure 2.233. Vertical panel delivery and installation. (Prototypic real-world application; image: Mathias Farnebo.)

Figure 2.234. Subsystems support efficient material handling and off-loading. (Prototypic real-world application; image: Terése Andersson.)

Figure 2.235. An overhead rail system allows the automation of on-site logistics for the handling of the panels. (Prototypic real-world application; image: Terése Andersson.)

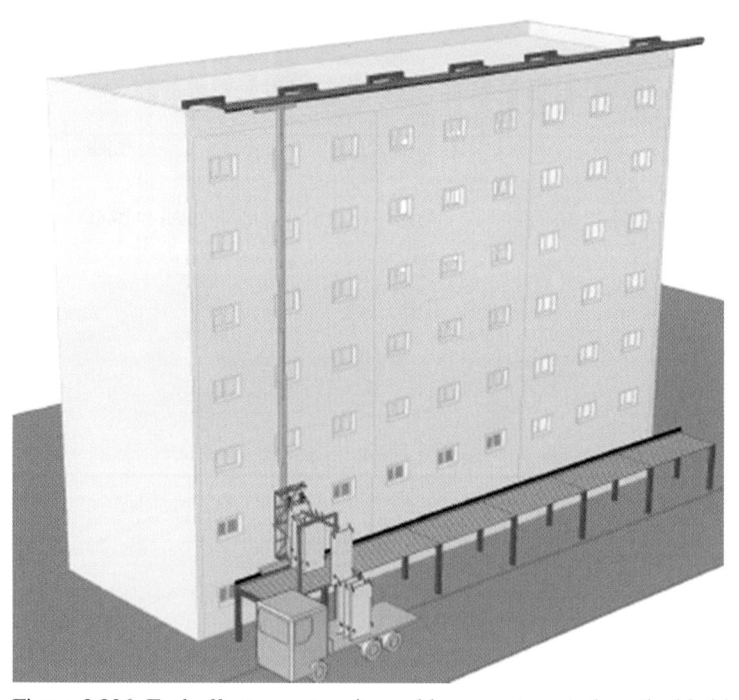

Figure 2.236. End-effector system is used by an automated vertical bridge crane. We can see the end-effector system elevating the facade component. (Image: Iturralde, Linner, and Bock, Chair of Building Realization and Robotics, Technische Universität München; see also Iturralde et al. 2015-2.)

of Germany and aims at both speeding up material logistics and reducing human labour input.

System 114: Robotic Installation Systems for Facade Panel Installation in Building Renovation

Developer: Chair of Building Realization and Robotics, Technische Universität München, Munich, Germany; Kepa Iturralde, Dr. Thomas Linner, and Prof. Thomas Bock

The objective of this research is to develop a robotic system that is suitable for the installation of panels onto existing buildings to improve the energetic performance of the envelopes. This early stage research has focussed on two main issues. One is the development of robotic support bodies (Iturralde et al. 2015-1). Second, there has been conception of the end-effectors (Iturralde et al. 2015-2). The participants of this research have considered that some support bodies will adapt better to different building typologies. For instance, a support body based on an aerial work platform is more suitable for low-rise buildings, whereas a cable-driven robot would be appropriate for a high-rise building. In this situation, the developers have taken into account that the end-effector system must be reconfigurable to any of the support bodies. The end-effector system must perform various tasks, but basically it will consist of the fixation of a connector and the posterior clipping of the panel into it. For that purpose, once the support body is placed in position, one of the end-effectors should drill holes. After that, another end-effector should place the connector and insert the fasteners in the holes. Finally, another end-effector should pick up the panel and place it in position. Currently this end-effector system is being developed. In the next stages of the research, the controlling system will be developed, as well as the robot-oriented design of the facade panel. Besides, within the EU-funded research project BERTIM (H2020 Research & Innovation Program under Grant Agreement No. 636984; http://www.bertim.eu) this approach will be proposed and contrasted with the rest of the partners, especially with the industry partners that install timber-based panels onto existing envelopes. Figures 2.236 and 2.237 show different types of approaches for robotic facade panel installation.

2.17 Tile Setting and Floor Finishing Robots

Buildings of all types are often clad with tiles made resistant to specific climate and weather conditions. In many countries, buildings such as single-family dwellings, factories, offices, and high-rise buildings are often equipped with tiles. Tiles are relatively small elements compared to the total surface area of a building, and huge amounts of tiles must be laid in the same, repetitive manner involving tile logistics, mortar application, and tile positioning. The large number of identical elements, the repetitiveness of operations, and the fact that facades are generally difficult to access makes the use of automated systems potentially rewarding. Tile-setting robots demonstrate that accuracy can be enhanced and that the even the laying of complex patterns can be accomplished without dramatically increasing the required man-hours or cost. Systems in this category include systems for the installation of tiles to vertical surfaces (walls, facades, etc.) as well as to horizontal surfaces (floors, etc.).

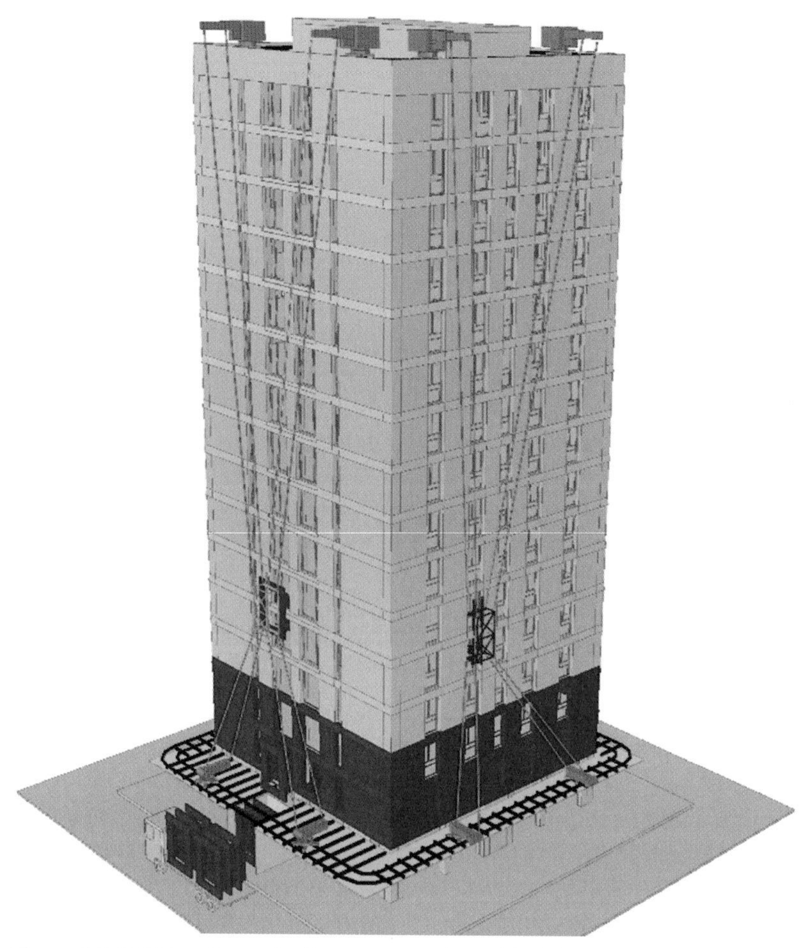

Figure 2.237. Using the end-effector system by cable-driven robots. (Image: Iturralde, Linner, and Bock, Chair of Building Realization and Robotics, Technische Universität München.)

All the identified systems work with a small intermediate storage that requires continuous refilling by a logistics solution.

Tile Setting and Floor Finishing Robots		
115	Robot for installation of tiles to facades	Hazama and Komatsu
116	Robot for floor tiling	Eindhoven University of Technology
117	Mobile robotic tiling machine	Future Cities Laboratory (FCL)/ROB Technologies AG.
118	Automated paving machine	Vanku B. V./ Tiger-Stone

System 115: Robot for the Installation of Tiles to Facades

Developer: Hazama Ando Corporation and Komatsu

The system automates the process of installation of tiles to the exterior walls of buildings (Figures 2.238 and 2.239). The system consists of the main tile setting

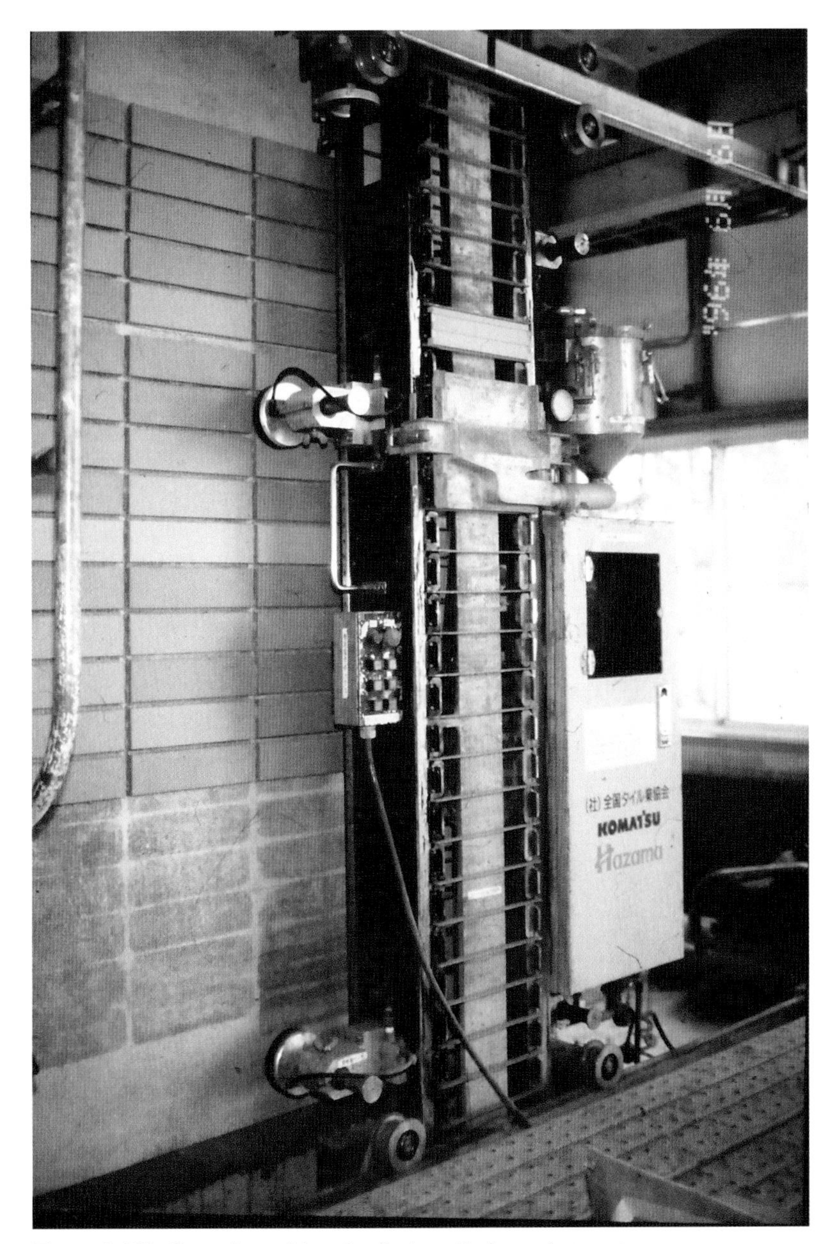

Figure 2.238. Overview of facade tile installation robot system.

unit (with a mechanism to store tiles, store and apply mortar, and place the tiles in a sequence to the wall) and a gantry-type mechanism oriented along the wall that moves the main tile-setting unit vertically during tile setting. Once one vertical segment of tiles has been completed, the gantry-type mechanism moves the main tile-setting unit to the next vertical segment. The robotic end-effector places and presses the tiles in sequence after applying a layer of mortar on the back of each tile. The system can be operated by one worker, and can achieve an accuracy of ±1 millimetre. The robot can process tiles of a size of 227 × 60 millimetres and with thicknesses ranging from 8 to 15 millimetres. The preparation of the wall as well as the installation on the site must be done by human workers.

Figure 2.239. Finished tile surface.

For further information, see Council for Construction Robot Research (1999, pp. 226–227).

System 116: Robot for Paving Floors with Ceramic Tiles

Developer: A+ Innovations, Eindhoven University of Technology, Kranendonk Production Systems, Hilhorst Tegelwerken

The system (Figure 2.240) consists of five key subsystems: (1) a mobile platform that allows the system to move smoothly on tile floors, (2) a robotic manipulator, (3) an end-effector with suction devices to handle multiple tiles, (4) a temporary on-board tile storage area (that can be resupplied by another complementary designed resupply robot), and (5) an on-board control and power supply system. The temporary on-board tile storage area can be resupplied by another complementary designed mobile resupply robot that is equipped with tiles in a robotic tile sequencing station. The end-effector is able to pick up and place up to four tiles at once. The robot can conduct tile setting in a preprogrammed as well as in a fully automated as well as

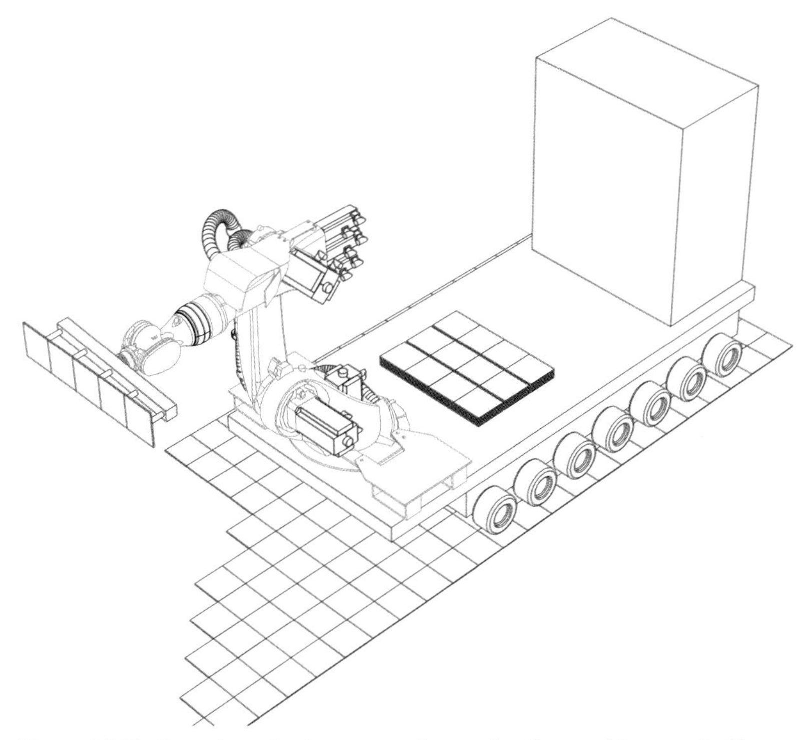

Figure 2.240. Overview of robot system for paving floors with ceramic tiles.

partly manually controlled manner. In addition, the robot is equipped with a laser sensor system that guides the robot within the working area through reference points. The robot was developed as part of a European funded research project with enterprises from the Netherlands, Germany, and Belgium.

For further information, see Lichtenberg and Segers (2000) and Lichtenberg (2003).

System 117: Mobile Robotic Tiling Machine

Developer: Future Cities Laboratory (FCL)/ROB Technologies AG

The robotic tiling machine (Figures 2.241–2.243) is a prototype of an STCR being developed by researchers from the Architecture and Digital Fabrication research module at the Future Cities Laboratory (FCL) of the Singapore-ETH Centre for Global Environmental Sustainability, in collaboration with its partner ROB Technologies AG, to automate the floor tiling process. Since early 2015, for the second phase of research and development, the robotic tiling machine project team at the Future Cities Laboratory (see also Future Cities Laboratory, 2015) has been collaborating with its new industry partners: SIKA Technology AG and LCS OPTIROC, with sponsorship from Bouygues Construction. Based on Universal Robot's UR5 (a relatively low-cost, collaborative lightweight robot) the system is able to work alongside human workers, laying floor tiles with high precision and consistent quality, while human workers can focus on higher value-added work such as cutting and placing special shaped or sized tiles (for the first generation of the machine) or

Figure 2.241. Automated adhesive application process. (The mobile robotic tiling machine is being developed by researchers from the Architecture and Digital Fabrication research module at the Future Cities Laboratory [FCL], in collaboration with its partner ROB Technologies AG.)

overseeing the operation of a group of robots. The prototype consists of a mobile base frame, a tile stack magazine (that allows tile storage), a lightweight robotic arm, and a custom end-effector that enables the picking and placing (setting) of a tile, as well as the handling of specialized tools that are in addition needed in tiling work such as the special nozzle extrusion tool for automated adhesive application that can be stored in and picked up from the mobile base frame. The adaptive control and planning software makes it possible to oversee the whole process with the data acquired from the sensory equipment and to generate a precisely laid tile pattern. According to the developers, once in the market, the robotic tiling machine could achieve up to a fourfold increase in tiling productivity in Singapore as it can work in a 24/7 mode in contrast to the manual tile laying process. Furthermore, it is expected that the robot reduces manual labour by 75% while quality and tile pattern complexity can be increased.

System 118: Automated Paving Machine

Developer: Vanku B. V./ Tiger-Stone

The Tiger-Stone automated paving machine (Figures 2.244–2.246) was developed and manufactured by a company from the Netherlands. The machine can automatically pave roads or pathways. The maximum width allowed for the road/pathway to be paved is 6 metres. The machine can be operated by one or two operators who basically only need to put the paving stones into the magazine on top of the machine. The machine automatically sorts the stones and arranges them

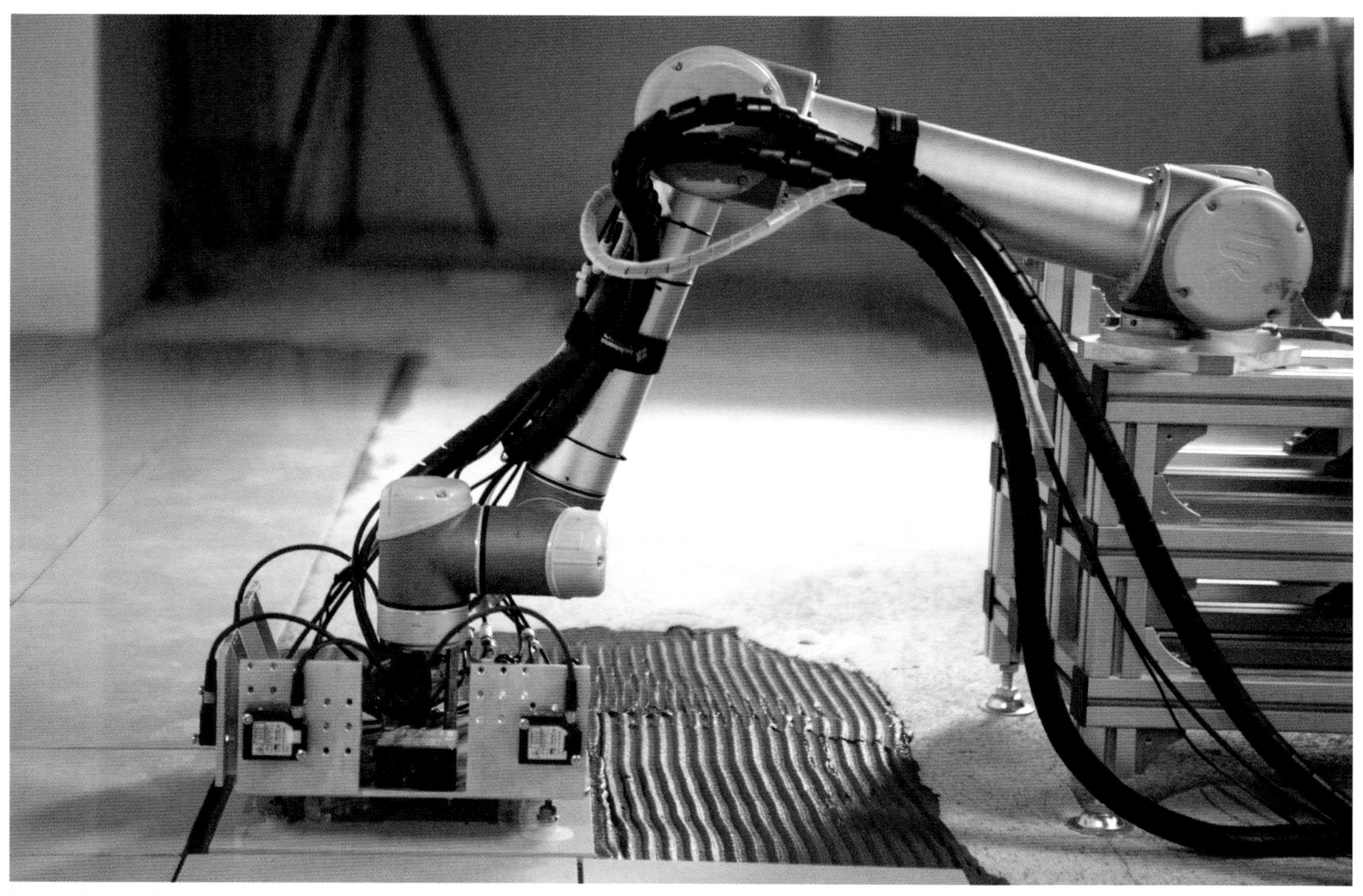

Figure 2.242. Tile laying process. (The mobile robotic tiling machine is being developed by researchers from the Architecture and Digital Fabrication research module at the Future Cities Laboratory [FCL], in collaboration with its partner ROB Technologies AG.)

Figure 2.243. Overview of mobile robotic tiling machine. (The machine is being developed by researchers from the Architecture and Digital Fabrication research module at the Future Cities Laboratory [FCL], in collaboration with its partner ROB Technologies AG.)

Figure 2.244. Unloading and deployment of the machine. (Image: Vanku B. V./Tiger-Stone.)

Figure 2.245. The machine can be operated by one or two operators. (Image: Vanku B. V./Tiger-Stone.)

Figure 2.246. The machine automatically sorts, arranges, and places the paving stones. (Image: Vanku B. V./Tiger-Stone.)

into predefined pattern. The maximum production capacity accounts for up to 500 m^2/day. The machine is compact and can be easily transported to the site by a van.

2.18 Facade Coating and Painting Robots

Facade painting robots were developed to simplify the painting of building facades, which, even with scaffolding, are often difficult to of access. Facades of high buildings, in particular, are difficult to paint or repaint during construction, as well as during operation. Facade painting robots have a particular advantage in keeping the quality constant. They usually have multiple spray nozzles operating in a synchronized mode. The nozzles are also usually encapsulated or housed within covered elements to prevent the escape of paint. The continuous painting quality is specifically controlled by the precise control of the spray nozzles, spraying speed, and spraying pressure. Another major advantage of painting robots is the fact that the workers are not exposed to harmful paint substances. STCRs for painting use different strategies to move along the facade such as suspended cage/gondola mechanisms, rail-guided mechanisms, and mechanisms allowing movement along the facade by vacuum or other adhesion technology.

The use of facade painting robots is not considered efficient below a facade area of 2000 square metres (Cousineau & Miura, 1998). Facade painting robots are therefore used primarily to paint large facades of high-rise buildings and larger types of commercial buildings. Facades to be painted are in general required to be rectangular, and to have no corners or profile structures that could hinder the operation of the robot. Furthermore, the design of the window frames as well as the amount and area covered by the windows impact the applicability and efficiency of facade painting robots. The operation speed of the systems of this category is allocated between 200 and 300 m^2/hour depending on the type of coating sprayed.

Facade Coating and Painting Robots		
119	Facade painting robot SB-Multi Coater	Shimizu
120	Facade coating and painting robot	Kumagai
121	TPR-02 facade painting robot	Taisei
122	Facade coating and painting robot	Urakami
123	Facade painting robot system	Kajima
124	Place Taisei logo on original image	Tokyu Construction Co., Ltd.
125	Facade painting robot	Kajima
126	Automatic balustrade and balcony facade painting system OSR-1	Shimizu
127	Automated coating robot for external walls	Takenaka

System 119: Facade Painting Robot SB-Multi Coater

Developer: Shimizu Corporation

The gondola-type robot (Figures 2.247 and 2.248) performs all the stages of painting of exterior walls and can spray the primer coat, middle coats, and finishing coat. The spraying/painting process is conducted automatically and supervised

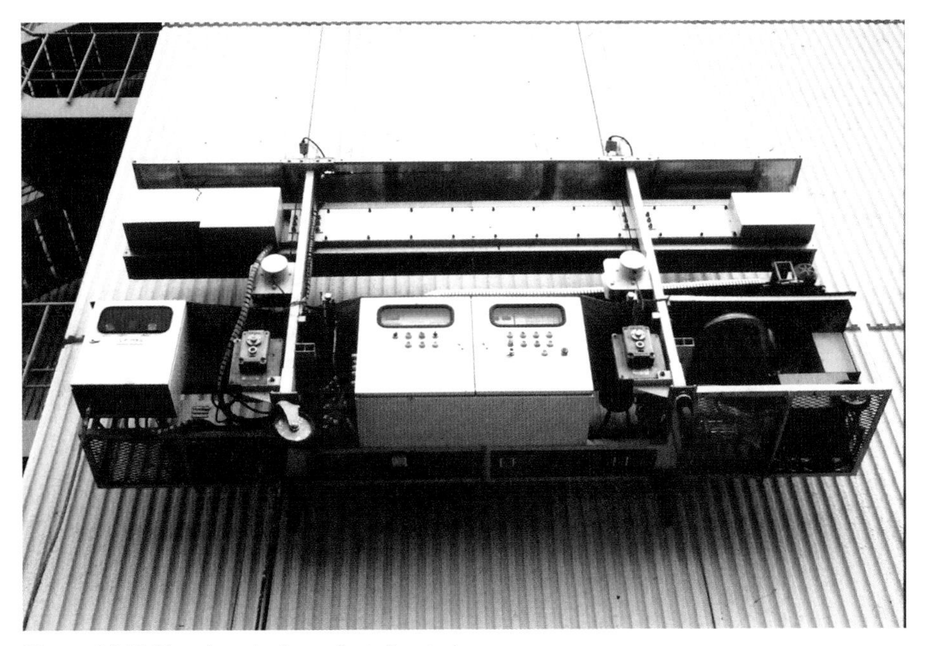

Figure 2.247. Facade painting robot, front view.

Figure 2.248. Side view of facade painting robot, side view.

through a control desk by a worker located on the ground. The robot consists of a main body to which two painting manipulators, sensors, and on-board control electronics are installed; a roof unit; and a pint feeding system located on the ground that pumps the paint though a hose to the main body. The main body is lifted and lowered along the building's facade surfaces automatically by the roof unit from which it is suspended. The robot is able to improve both the quality and the efficiency of the painting work as it allows precise computer control of the spraying process, ensuring a uniform and exact distribution of the coating or painting material on the wall surface.

System 120: Facade Coating and Painting Robot

Developer: Kumagai Gumi

The robot system (Figure 2.249) was developed to automate surface finishing tasks such as cleaning, polishing, surface preparation, and painting (grinding width: 0.30–0.33 m, painting width: 0.35–0.37 m, max. force: 80 kgf [grinding], 130 kgf [painting]). The robot consists of a self-driven mobile unit (dimensions: L 0.98 m × W 0.65 m × H 0.35 m; max. travel speed: 5–6 m/minute) equipped with a vacuum suction system that allows it to adhere to and climb vertically along the wall, an end-effector, a roof unit (a support system from which the self-driven mobile unit is suspended from the roof to provide backup support in the event of a failure of the suction system and for horizontal repositioning), and a vacuum support and paint/cleaning material supply system located on the ground (dimensions: L 0.20 m × W 3.00 m × H 2.00 m. The

Figure 2.249. Facade coating and painting robot. (Image: Kumagai Gumi.)

Figure 2.250. Robot for painting exterior wall system in action. (Image: Taisei.)

robot can be equipped with a variety of end-effectors including grinders, polishing tips, and spray painters. The robot is supervised by an operator on the ground using a camera system installed on the robot. The robot can navigate a variety of facade surfaces such as glass, steel, and concrete surfaces and is also capable of moving along surfaces with slight irregularities. The robot is also able to detect larger obstacles on the surface and to stop the operation automatically if necessary.

System 121: Robot for Painting Exterior Walls

Developer: Taisei Corporation

The robot for painting exterior walls (Figures 2.250–2.252) from Taisei is used for the automated application of paint to the facade of a high-rise building. The system allows for improved efficiency and accuracy as well as reduced risk as manual labour at extreme heights is not required. The system consists of three main components; a mobile, rail-guided robot unit (vertical movement along the facade; speed: 8 m/minute descending, 16 m/minute ascending) equipped with a mechanism that moves the actual panting end-effector, horizontally, a roof-mounted unit from which the robot unit is suspended (this unit also integrates a control cabin as well as tanks to store panting/coating material), and a paint supply component that transports paint from the roof carriage to the robot unit. The robot moves vertically along the face of the building along guiderails that were previously integrated into the building's facade. The painting end-effector of the robot is housed within a dedicated

Figure 2.251. Robot travelling along the faced through facade-integrated rails. (Image: Taisei.)

Figure 2.252. Overview of robot for painting exterior wall system. (Image: Taisei.)

compartment such that no excess paint escapes. Eight rotating airless-type spray guns of the end-effector apply paint in a precise and efficient manner. The thickness of the paint/coating applied can be accurately controlled and adjusted. A series of sensors allow the robot to paint accurately around facade elements such as windows. The system is supervised by a single operator through the control cabin that is part of the rooftop unit. Paint/coating material is supplied from the tanks of the rooftop unit through a hose to the main unit. After each vertical section is completed, the robot is relocated through the roof unit (that can travel horizontally along the rooftop) to the next vertical section. The robot is able to achieve a painting rate of about 100 m^2/hour.

For further information, see Council for Construction Robot Research (1999; pp. 248–249) and Takeno et al. (1989).

System 122: Facade Coating and Painting Robot

Developer: Urakami

The facade coating and painting robot (Figure 2.253) from Urakami was designed to polish, coat, and paint large building surfaces. It combines a gondola-type suspension strategy with a vacuum-based system that attaches the robot to the wall. The robot system consists of the main robot unit that contains the end-effectors, which is held by a vacuum system on the wall; a roof-mounted unit that contains the activators for the gondola type positioning system; and a support unit installed on the ground for storing and supplying painting and coating material through a hose system to the main robot unit. The working width along the axis of vertical movement is 37 centimetres for painting/coating and 33 centimetres for polishing. The maximum processing speed of the system is 6 m/minute. The dimensions of the main robot unit are L 98 cm × W 65 cm × H 35 cm.

Figure 2.253. Facade coating and painting robot system in action. (Image: Urakami/Construction Robot System Catalog in Japan, 1999, pp. 322–323.)

For further information, see Council for Construction Robot Research (1999, pp. 322–323).

System 123: Facade Painting Robot System

Developer: Kajima Corporation

In contrast to other painting robot systems, this robot system is not suspended from the top of the building; but rather, it is based on a gantry-type kinematic structure oriented horizontally along the wall (Figure 2.254). The robot system is able to coat and/or paint large surfaces of a building. It consists of the gantry-type positioning system, a covered spraying chamber, a robotic manipulator placed inside the spraying chamber, a system for storing and supplying the paint (installed on the ground and able to pump the coating/painting material to the spraying chamber), and the spraying gun end-effector. The gantry-type positioning system allows the actual spraying system (housed in the covered spraying chamber) to be moved at relatively high speeds all over the facade. The gantry-type positioning system consists of a vertical mast (installed between temporary installed rails that are fixed between the roof and a mobile, robotic carriage on the ground of the building). The mast can be guided by the rails and the motorized carriage moved horizontally along the wall. Along the mast the painting chamber can be moved in the vertical direction. Inside the spraying chamber, the spraying gun end-effector is moved by a rail-system–based manipulation system back and forth (z-direction), left and right (x-direction), and up and down (y-direction). Once positioned by the gantry-type positioning system this system can thus cover a work area of 4 × 1 metres. When one work area is fully processed the system is repositioned by the gantry-type positioning system. The work procedure always starts at the top of the wall and the spraying chamber is repositioned until

Figure 2.254. Facade painting robot system. (Image: Kajima Corporation.)

Figure 2.255. System equipped with end-effector for facade painting. (Image: Tokyu Construction Co., Ltd.)

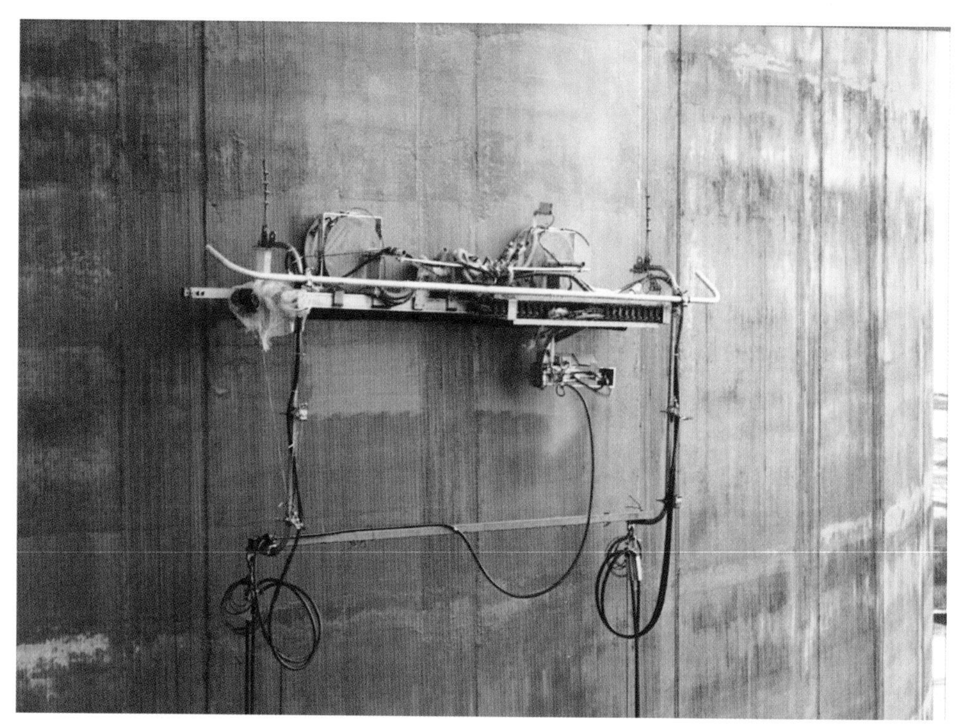

Figure 2.256. System equipped with an end-effector for high-pressure water cleaning of the silo surface. (Image: Tokyu Construction Co., Ltd.)

one vertical wall/faced segment is processed. Following this the next vertical (4 m broad) wall segment is processed. A sensor system detects irregularities in the wall and thus allows the system also to paint walls containing windows or others design features. The system is able to perform coating/painting with a processing rate of up to 200 m^2/hour and special textures spraying with a processing rate of up to 100 m^2/hour.

For further information, see Terauchi et al. (1993).

System 124: Facade Diagnostics, Cleaning, and Painting Robot

Developer: Tokyu Construction Co., Ltd.

This robot can not only be used for painting (Figure 2.255) but can also be equipped with end-effectors for cleaning (Figure 2.256) and as a diagnostic of wall tile conditions. The robot can be installed on any large building surface and the main body is suspended along two wire ropes allowing it to move vertically along the building's surface with the help of an automated winch. Once the robot is brought into the correct position by the winch, it is temporarily fixed to the wall by a suction system. Once one vertical lane of the building is painted, the robot is moved automatically to the next lane. By changing the attached end-effector's head, the robot can switch between the previously mentioned tasks. Using the painting end-effector, which contains a mechanism that moves a painting head over the surface similar to the mechanism of an ink-jet printer, the robot can even print patterns or images in one or even several colours. Waste of paint is minimized by exactly dosed paint injection and paint waste absorption. The system can be operated automatically following a program or remotely using a remote control unit.

Figure 2.257. Overview of facade painting robot system.

System 125: Facade Painting Robot

Developer: Kajima Corporation

This painting robot belongs to the gondola type category, as the main unit that houses the painting end-effectors is moved along the facade vertically and horizontally by a mobile activating unit temporarily installed on the roof of the building (Figures 2.257 and 2.258). The system is able to execute all stages of the painting process for the exterior wall through its ability to spray the primer coat, middle coat,

Figure 2.258. Facade painting robot system in action.

and finishing coat. It consists of a unit located on the roof, the suspended main unit (dimensions: L 5.4 m × W 0.8 m × H 1.5 m; weight: 350 kg), a rail system–based robotic manipulator that moves the spraying gun end-effectors, and a material supply system located on the ground supplying material to the main unit through a hose. The spraying chamber and the spraying gun end-effector are moved in by the rail-system–based manipulation system back and forth (z-direction), left and right (x-direction), and up and down (y-direction). A sensor system (surface detection sensors synchronized with the rail system–based robotic manipulator and the spraying gun) detects irregularities in the wall and thus allows the system also to paint walls containing windows or others design features. The system is able to perform coating/painting with a processing rate of up to 290 m^2/hour. The spraying and painting processes are conducted automatically and are supervised by a worker located on the ground.

System 126: Automatic Balustrade and Balcony Facade Painting System OSR-1

Developer: Shimizu Corporation

The OSR-1 from Shimizu is used for balustrade and outer wall finishing work. The system operates from the balcony on which it is temporarily deployed and eliminates the need for repetitive manual work to be completed from scaffolding on the exterior of buildings (Figures 2.259 and 2.260). The system consists of a travelling device (mobile platform placed on the floor of the balcony on which the system is deployed, a manipulator (consisting of a horizontal link and a vertical link) that embraces the balcony and positions the end-effector on the outside of the balustrade, a spray gun end-effector, and guiding mechanism that helps the robot to be guided

Figure 2.259. OSR 1 painting the outside of a balcony. (Image: Shimizu Corporation.)

Figure 2.260. The robot moves along the handrail of the balcony, wall, or balustrade. (Image: Shimizu Corporation.)

along the balustrade. The end-effector moves up and down along the vertical link of the manipulator to apply the paint. A separate air compressor system (that supplies air pressure to the spray gun) as well as a tank with the painting/coating material is installed to and carried along with the robot by the mobile platform. The robot moves along the handrail of the balcony, wall, or balustrade or any other type of temporarily installed guiding rail. For uneven sections of a wall, the robot can rotate its horizontal link, follow the unevenness, and still apply the paint as intended. Further specifications of the system:

- Painting rate: 80 m^2/day
- Max. travel speed: 20 m/min
- Arms: Right–left turning: ± 135 degrees at 10 degrees/s; up–down turning: 0–95 degrees at 15 degrees/s; in–out: 0.25 m at 0.15 m/s; up–down traverse: 1.80 m at 0.900 m/s
- Dimensions: L 0.65 m × W 0.45 m × H 1.565 m
- Weight: 223 kg
- Power source: AC 100V, 50/60 Hz. Pneumatics 7 kgf/cm^2

Figure 2.261. Facade coating robot. (Image: Takenaka Corporation.)

System 127: Automated Coating Robot for External Walls

Developer: Takenaka Corporation

The conventional method of painting facades of buildings can be time and labour intensive, and it requires skilled labour to achieve results of satisfactory quality. The automated coating robot was developed by Takenaka to improve work efficiency by automating the painting/coating process. The system consists of a mobile crane with a vertical outrigger, a main unit with a coating/painting end-effector, and a material supply system installed on the mobile crane. The main unit travels vertically (up and down) along the outrigger of the mobile crane (Figure 2.261), which is positioned in front of the wall/facade to be processed. The end-effector (Figure 2.262) is located at the bottom of the main unit and uses a wide-angled, multinozzle system to ensure uniform painting coverage along the facade. The system can be adjusted to handle various paint exterior wall and paint types.

2.19 Humanoid Construction Robots

The field of humanoid robotics can be considered as one of the technically most complex fields of robotics cause by challenges associated with humanoid robots such as the usually complex kinematic structures, the many degrees of freedom, the bipedal locomotion mechanisms, and last but not least the demand for safe interaction with human beings. As a result of this and the enormous cost of the hardware and software implemented in humanoid robots as well as the – as the DARPA Robotics Challenge recently showed – still difficult and error-prone autonomous or partly autonomous operation of humanoid robots, up-to-date humanoid robots are not yet reality in any of the long envisioned applications fields such as care,

Figure 2.262. Robotic end-effector for facade coating. (Image: Takenaka Corporation.)

manufacturing, or construction. However, companies and universities continue to invest in the development and exploration of humanoid robotics because of its frontier nature, the ultimate promises this approach could realize, and the fact that indeed many technologies developed in the context of humanoid robotics and robotic challenges can later be used in adapted form in other robotic systems or machines. Apart from this, particularly in the last decade, a number of companies commercialized their humanoid robots successfully at least as research platforms sold to universities and other research institutes (Kawada is one of these companies) and very recently in the form of significantly simplified systems, even as manufacturing or service ready commercial versions (Rethink Robotics). Despite the aforementioned challenges, humanoid robots would present a very smart approach towards construction automation as they would be fully compatible with existing machines (instead of a human being, they could automate any equipment on demand just by operating it), could directly and naturally communicate with human beings, assist on the construction site with carrying/handling material, collaborative installing plasterboard panels, and assemble steel profiles. Executives of major Japanese contractors have confirmed to the authors that they see, in particular, a huge potential in breakthroughs in humanoid robotics as an alternative to other approaches outlined in this volume.

Humanoid Construction Robots		
128	Baxter/Sawyer	Rethink Robotics
129	CHIMP – CMU highly intelligent mobile platform	DARPA Robotics Challenge (DRC) Tartan Recue Team at Carnegie Mellon University
130	HRP robot series	Kawada, AIST, NEDO, MSTC, METI, Yasukawa, Shimizu Corporation, Honda

System 128: HRP Robot Series

Developer: Kawada Robotics Corporation, AIST, NEDO, MSTC, METI, Yasukawa, Shimizu Corporation, Honda

The development of the HRP series involved several institutions and companies, with two key players (Kawada, Shimizu) coming from the construction industry. Kawada is a construction company that today operates its own robotics division and that is producing and commercially marketing humanoid robots. One of the main purposes of the HRP robot series development was to provide a humanoid robot that can assist and work alongside humans on construction sites. In 2002 the HRP-2 (height: 1.54 m; weight: 58 kg; DOF: 30) was completed, in 2007 the HRP-3 (height: 1.6 m; weight: 68 kg; DOF: 42), and in 2010 the HRP-4 (height: 1.51 m; weight: 39 kg; DOF: 34; Figure 2.263). From version to version, features such as operation time, design, weight, and collaborative ability were continuously improved. The HRP-4 can work beyond 120 minutes powered by an on-board battery pack and follows a lightweight robot concept applied to a humanoid robot. An open source software development platform (OpenRTM-aist) is available so that purchasers of the robot can develop their own applications. Today the HRP-4 platform is popular among research institutes around the world as an experimentation platform and can be purchased from Kawada for a price of €350.000–450.000 (depending on the equipment with sensors, end-effectors, etc.). The robot is still produced only in very small quantity, which makes parts and production costly. However, it can be assumed that mass production will in the future be able to reduce the price considerably. So far, it has been demonstrated that the HRP robots can assist on the construction site with carrying/handling material (Figure 2.264), collaboratively install plasterboard panels, assemble steel profiles (Figure 2.265), and drive/operate construction equipment as diggers or forklifts (Figure 2.266). Executives of major Japanese contractors have confirmed to the authors that they see, in particular, a huge potential in the last feature mentioned, which would make it possible to turn existing, conventional construction equipment into autonomous machines on demand.

Figure 2.263. HRP-4 – the latest HRP version. (Image: © Kawada Robotics Corporation, AIST.)

Figure 2.264. HRP-2 assisting in panel handling and installation. (Image: © AIST, Kawada Robotics Corporation.)

Figure 2.265. HRP-3 conducting assembly work on the site. (Image: © AIST, Kawada Robotics Corporation.)

System 129: CHIMP – CMU Highly Intelligent Mobile Platform

Developer: DARPA Robotics Challenge (DRC) Tartan Rescue Team at Carnegie Mellon University, National Robotics Engineering Center (NREC). The material about CHIMP presented below is based on work supported by the US Navy SPAWARS Center under Contract No. N65236-C-3886. Any opinions, findings, and conclusions or recommendations expressed in this material are those of the authors and do not necessarily reflect the views of the US Navy SPAWARS Center.

Figure 2.266. HRP-1S controlling a construction machine. (Image: Prof. Maeda.)

Figure 2.267. CHIMP can handle tools such as screwdrivers. (Image: Photo Courtesy of the National Robotics Engineering Center, Carnegie Mellon University.)

CHIMP was developed in the context of the DARPA Robotics Challenge by Carnegie Mellon University's National Robotics Engineering Center (NREC). As part of the challenge, the individual teams participating had to develop robots that are able to conduct tasks in unstructured and dangerous environments and use cases (Figures 2.267–2.271) such as the Fukushima nuclear power plant incident. However, the robots and functionality that were developed in that context also represent interesting approaches and applications for construction and maintenance tasks. CHIMP in particular is suitable for construction-related tasks because its kinematic

Figure 2.268. To go over obstacles CHIMP can transform into a robot with four legs that can use its legs both for walking as well as tracks integrated in the legs for movement. (Image: Photo courtesy of the National Robotics Engineering Center, Carnegie Mellon University.)

Figure 2.269. CHIMP can be used for utility handling and maintenance. (Photo courtesy of the National Robotics Engineering Center, Carnegie Mellon University.)

Figure 2.270. CHIMP using tools for opening a shaft. (Photo courtesy of the National Robotics Engineering Center, Carnegie Mellon University.)

Figure 2.271. CHIMP handling wood parts in an unstructured environment. (Photo courtesy of the National Robotics Engineering Center, Carnegie Mellon University.)

structure allows flexibility (a multitude of tasks can be carried out by the robot) and robust mobility in a variety of terrains. The robot is inspired by human and ape bodies and integrates mobility elements with tracks. Depending on the task and terrain, the robot can move forward on two legs by tracks so that its upper body is in an upright position and its two arms can be used to conduct certain tasks. Furthermore, to go over obstacles CHIMP can transform into a robot with four legs that can use its legs both for walking as well as tracks integrated in the legs for movement. CHIMP has about 30 DOF and its control strategy aims at receiving and processing high-level commands and conduct low-level autonomy.

System 130: Baxter/Sawyer

Developer: Rethink Robotics

The person behind Rethink Robotics is Rodney Brooks, who is a specialist in developing highly affordable robots. Similar to iRobot's Roomba service robots (Rodney Brooks was a cofounder of iRobot), Rethink Robotics' systems combines intelligently reduced functionality with innovative features, thus allowing the robot to be sold at low cost and nevertheless meets the needs of a large number of standard work tasks. One of the key features of Rethink Robotics robots is that they are collaborative robots with compliant joints with series elastic activators, force sensing, and force control that eliminate the need for safety cages and allow them to cooperate directly with human beings. Furthermore, the robots do not necessarily need programming but workers can intuitively teach the robot certain movements. Baxter is equipped with two 7-DOF arms, a sonar and a front camera, vision-guided object detection features, changeable end-effectors, and a mobile base. Sawyer is equipped with only one 7-DOF arm but its special kinematic structure makes it a little more dexterous and accurate. With a price of about €20,000–30,000 (depending on the software used and equipment with end-effectors and other add-ons) Baxter is (in contrast to other lightweight, collaborative robots on the market) relatively inexpensive. The low price, the collaborative features, and the teaching capability make Baxter and Sawyer highly interesting for construction to small and medium-sized enterprises (in particular, assisting craftsmen and construction workers during

Figure 2.272. Rethink Robotic's Baxter as a cooperative type of robot could assist craftsmen and construction workers in many ways in collaborative scenarios in the future. (Image: Walnut Creek Planing.)

assembly tasks; see, e.g., Figure 2.272). Moreover, the low weight of the robots (Baxter: 75 kg; Sawyer: 19 kg) and the fact that they require no separate power and control box, as many standard industrial manipulators do, makes them highly mobile.

2.20 Exoskeletons, Wearable Robots, and Assistive Devices

Exoskeletons, wearable robots, and other assistive, cooperative robots and devices allow, through direct cooperation between human beings and the robot system or device, to combine the flexibility and intelligence of human beings with the strength, speed, precision, and endurance of machines and robots. The approach, on the one hand, makes it possible to circumvent many of the challenges associated with other types of robots (e.g., no need for full automation or complex locomotion or navigation strategies and thus definitely much easier to develop and implement than humanoid robots) and on the other hand introduces new challenges in terms of biosensors, human–robot interaction, and control algorithms required. Full-body exoskeletons (e.g., HAL from Cyberdyne) that embrace the whole human body are the most capable but also most complex form of wearable robots. In particular for manufacturing or construction purposes such exoskeletons may be equipped with additional handling devices or mini cranes (as, e.g., in the case of the DSME exoskeleton, FORTIS exoskeleton, HEXAR-PL). To bring such systems faster from research stages to the market recently tradeoffs between performance and the system complexity (e.g., in terms of the amount of joints covered by the exoskeleton) are considered leading currently to extensive research in the field of partial (e.g., Honda's walking assist device, HAL for lower limb, HAL light labour and lumbar support), one joint (HAL single-joint device), and unpowered exoskeletons (without activators; e.g., FORTIS exoskeleton). Indeed for many applications and tasks in construction full-body exoskeletons can be considered as overengineered, and simpler task-adapted solutions might represent less expensive and more robust solutions with a better usability. A novel approach that might be interesting in particular in the context of facade element installation or interior finishing assembly is MIT's supernumerary robotic limbs (SRLs; worn by the user as a back-pack like systems) that provide a worker with additional limbs that can operate independently from the arms and legs of the worker. Furthermore, part of the robot categories outlined in this chapter are, in contrast to exoskeletons, much simpler and currently more marketable cooperative robots (e.g., Kuka's Intelligent Industrial Work Assistant or the robots of the company Universal Robots) as well as the numerous recently developed smart glass applications.

Exoskeletons, Wearable Robots, and Assistive Devices		
131	Exoskeleton for handling heavy steel elements	Daewoo Shipbuilding and Marine Engineering (DSME)
132	HAL® Series: hybrid assistive limb technology	CYBERDYNE Inc.
133	FORTIS exoskeleton	Lockheed Martin, US National Centre for Manufacturing Science
134	Lower extremity exoskeleton robot for concrete placing (HEXAR-PL)	Department of Robot Engineering, Hanyang University
135	Walking assist device with bodyweight support system	Honda Motor
136	Supernumerary robotic limbs (SRLs)	d'Arabeloff Lab, MIT, PI: Professor H. Harry Asada
137–139	Smart glasses • Vuzix M200AR waveguide HD device • Microsoft HoloLens • Google Glasses	various developers; e.g., Vuzix, Google, Microsoft, etc.
140	Affordable and modular cooperative robots	Universal Robots

System 131: Exoskeleton for Handling Heavy Steel Elements

Developer: Daewoo Shipbuilding and Marine Engineering (DSME)

DSME's exoskeleton is made from carbon, aluminium, and steel, weighs 28 kilograms, and is fully self-supporting, meaning that the worker wearing the suit is freed from the weight of the exoskeleton itself. It is powered by hydraulic and electric activators that allow the exoskeleton a lifting capacity of up to 30 kilograms (which is beyond what the human worker has to handle). Moreover, the exoskeleton can be equipped with a variety of task-specific add-on frames that turn the exoskeleton into a mini crane for material handling (Figure 2.273). At the back of the exoskeleton a battery pack is installed that allows for up to 3 hours of operational time. The goal of the company is to improve the capability of the exoskeleton so that in the future it will be able to take over loads of more than 100 kilograms. Although Daewoo developed this exoskeleton in the context of shipbuilding, its use in the near future in other divisions of Daewoo, such as its construction division (Daewoo E&C), is highly likely. It is common in the Japanese and Korean industries, where companies usually hold divisions of many industry branches, that, for example, welding robots

Figure 2.273. Daewoo's exoskeleton being tested in the context of handling heavy steel elements. (Image: Daewoo/Daewoo Shipbuilding and Marine Engineering [DSME].)

developed in the shipbuilding division are transferred to the construction division later on.

For further information, see IEE Spectrum (2015) and DSME (2015).

System 132: HAL® Series: Hybrid Assistive Limb Technology

Developer: CYBERDYNE Inc.

Description: The world's first cyborg-type robot, Robot Suit HAL, which can improve, support, and expand a human worker's physical functions, has been

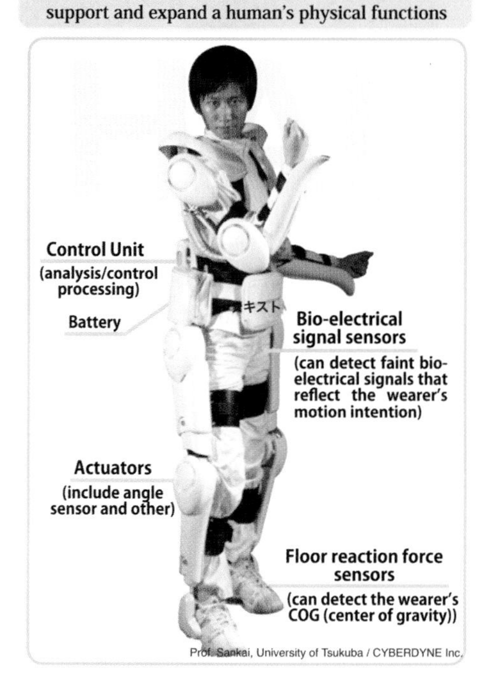

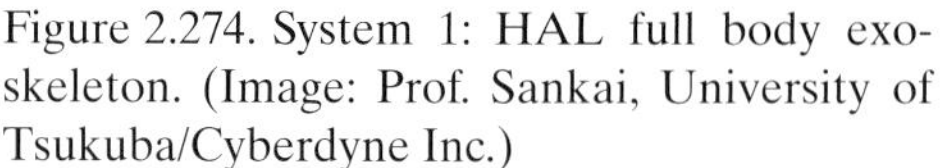
Figure 2.274. System 1: HAL full body exoskeleton. (Image: Prof. Sankai, University of Tsukuba/Cyberdyne Inc.)

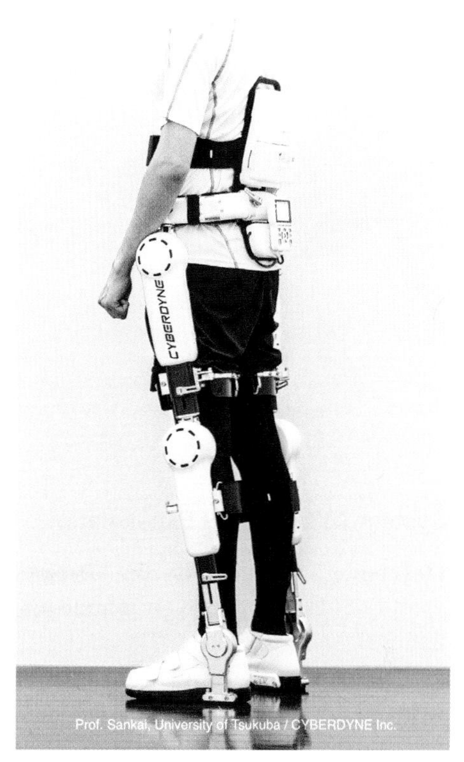

Figure 2.275. System 2: HAL for lower limb. (Image: Prof. Sankai, University of Tsukuba/Cyberdyne Inc.)

developed by Prof. Sankai at University of Tsukuba beginning in 1991, based on the new academic field "Cybernics: Fusion of Human, Robot and Information Systems". Later on, HAL was transferred to a company called CYBERDYNE (founded 2004) for commercialization on a large-scale and worldwide level. The key technology of the suits are biosensors that detect motion command signals that go from the brain through neurons into the body parts intended to be moved. The detected signals are used as the basis for a proactive, assistive control of human motions. Over time various system variations were developed. The HAL fully body exoskeleton (system 1; Figure 2.274) was developed for a prototype robot exhibition in the 2005 World EXPO, Japan. Depending on the use case and tasks to be accomplished, the HAL for lower limb (system 2; Figure 2.275) or a single-joint HAL (system 3; Figure 2.276) may represent a suitable solution for medical/therapeutic and healthcare fields. Recently, to make HAL technology improve workers' environments on construction sites the HAL heavy labour and lumbar support (system 4; Figure 2.277) for workers was developed. In general, the weight capacity has to be below 40% of body weight, for example, approximately 34 kilograms (body weight 85 kg). HAL technology originally was tested and developed in the context of medical applications. Over time additional application fields emerged and in particular systems 3–4 open up new possibilities in fields such as disaster response, factory work, and construction.

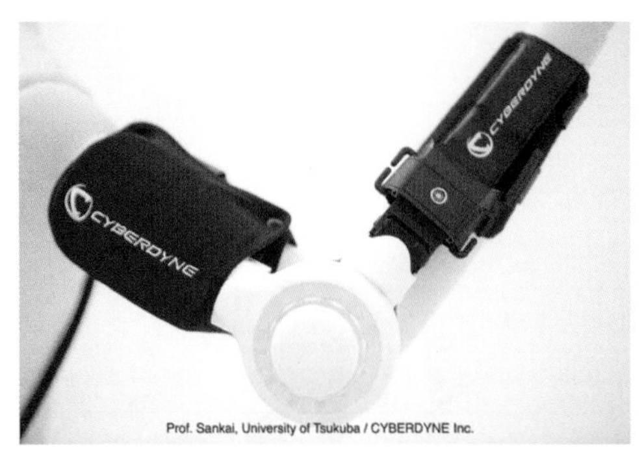

Figure 2.276. System 3: HAL single-joint device. (Image: Prof. Sankai, University of Tsukuba/Cyberdyne Inc.)

Figure 2.277. System 4: HAL light labour and lumbar support. (Image: Prof. Sankai, University of Tsukuba/Cyberdyne Inc.)

System 133: FORTIS Exoskeleton

Developer: Lockheed Martin, US National Centre for Manufacturing Science

FORTIS is a relatively simple, unpowered (containing no active motors), and lightweight exoskeleton. The idea is that heavy handheld construction devices or end-effectors such as welding tools, sandblasters, and grinding machines can be fixed to an additional mechanical arm extending from the exoskeleton (Figures 2.278 and 2.279). This arm then allows these devices or end-effectors to be fixed conveniently in certain positions and their load transferred to the ground over the exoskeletons. Furthermore, a device on the back of the exoskeleton allows balance weights to be added, which in addition makes it possible to counterbalance the weight of the tool fixed to the extended arm. Workers can thus endure certain positions longer and without fatigue, and also usually awkward positions become ergonomically feasible.

Figure 2.278. FORTIS exoskeleton equipped with a welding tool. (Image: Photo Courtesy of Lockheed Martin. Copyright 2015.)

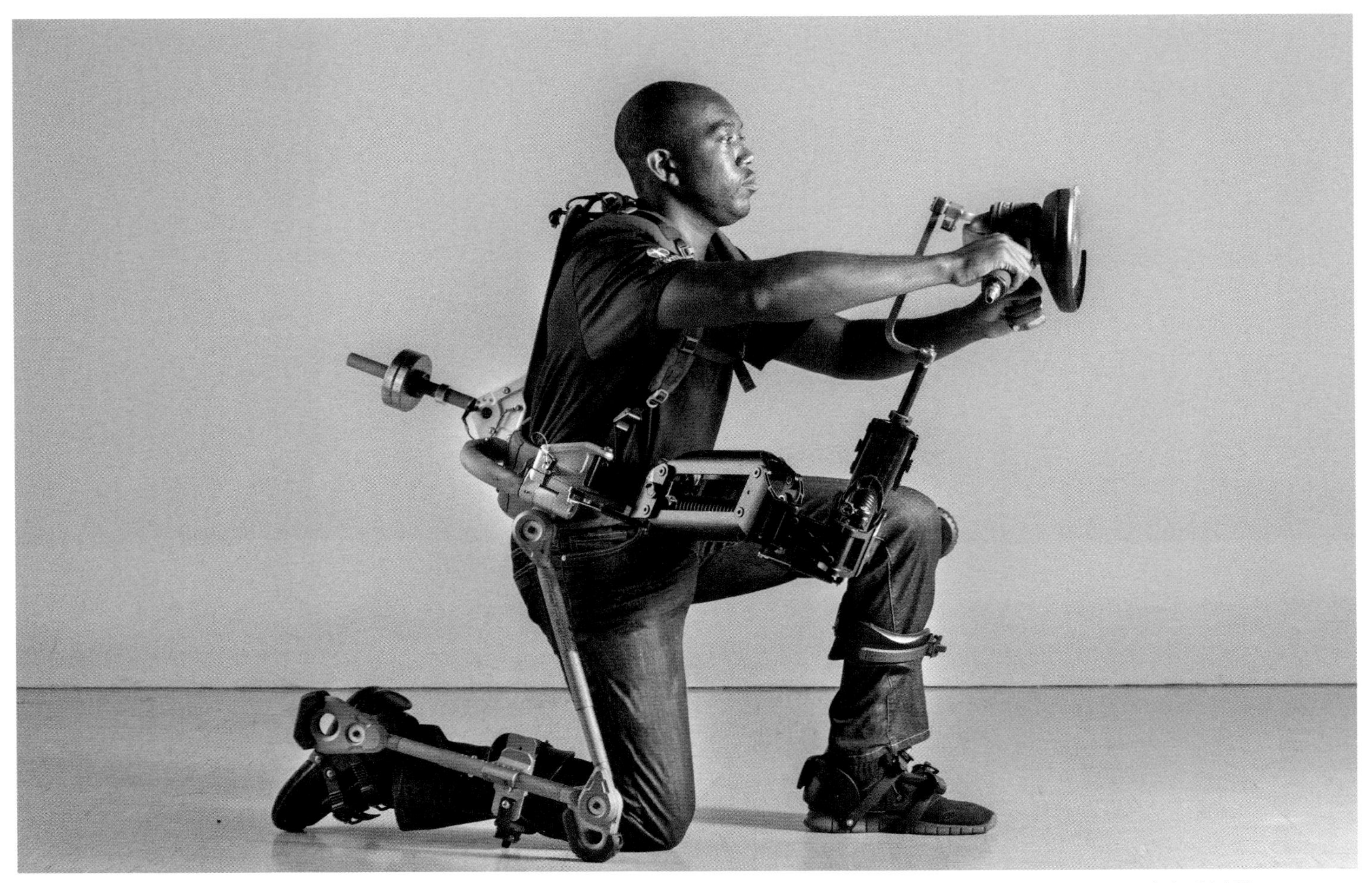

Figure 2.279. FORTIS exoskeleton equipped with a grinding tool. (Image: Photo Courtesy of Lockheed Martin. Copyright 2015.)

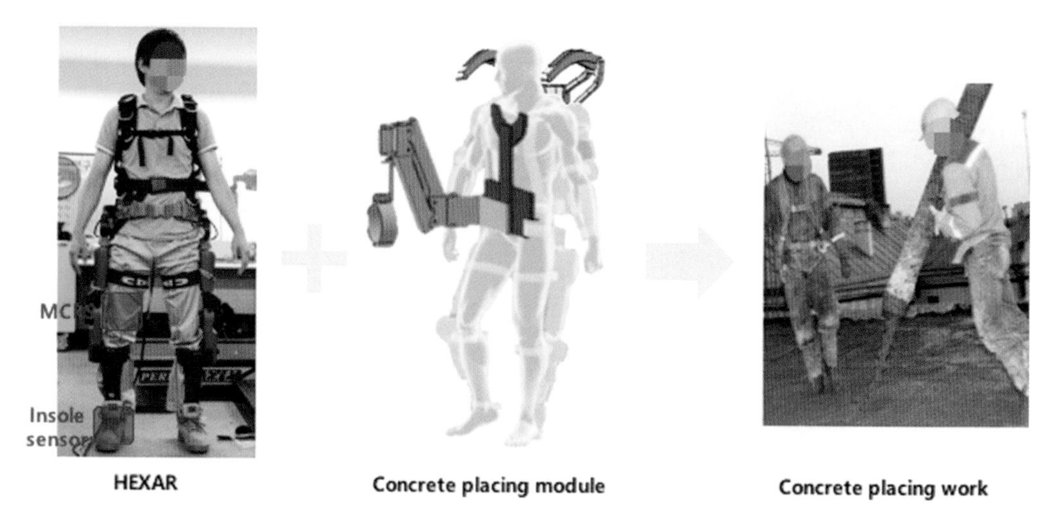

Figure 2.280. Design concept of the HEXAR-PL. (Image: Department of Robot Engineering, Hanyang University.)

The FORTIS exoskeleton is considered by Lockheed Martin as a transitional device to more advanced exoskeletons with active motors. Although the device is not for sale yet, it is likely that this exoskeleton (as it contains no active motors or sensors) will be available for a price far below that of other exoskeletons on the market.

System 134: Lower Extremity Exoskeleton Robot for Concrete Placing (HEXAR-PL)

Developer: Department of Robot Engineering, Hanyang University, Ansan Korea, Prof. Chang Soo Han, PhD, Seunghoon Lee, PhD, Wan Soo Kim, PhD, Hee Don Lee, Myeong Su Gil, and Dong Hwan Lim

The lower extremity exoskeleton system HEXAR (Figures 2.280–2.282) increases the muscle strength of the wearer during movement and work with heavy

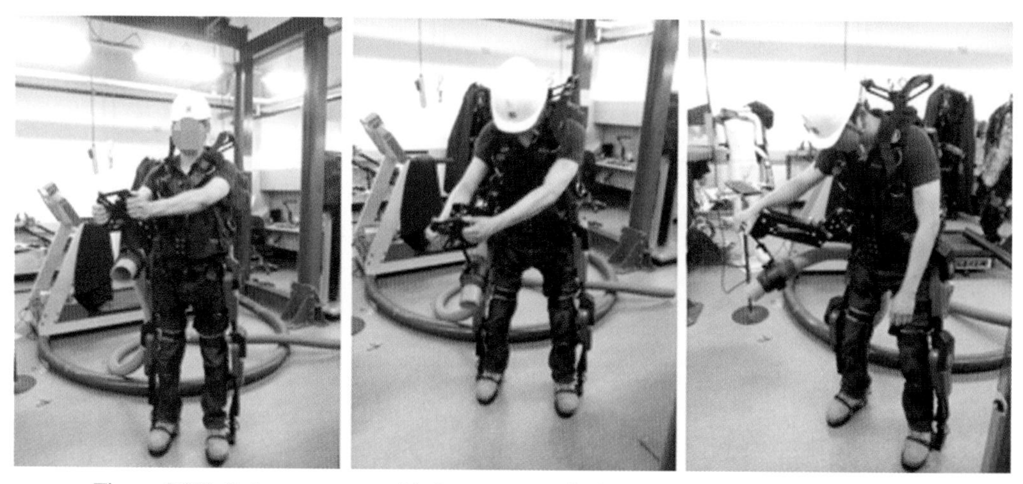

Figure 2.281. Laboratory test with the concrete placing module. (Image: Department of Robot Engineering, Hanyang University.)

Figure 2.282. Concrete placement filed test with the HEXAR-PL. (Image: Department of Robot Engineering, Hanyang University.)

materials. Work types in which dependence on human labour is high owing to demanded flexibility and where at the same time heavy materials have to be handled are suitable for being enriched with exoskeleton technology. For load carrying, the HEXAR consists of a lower exoskeleton and a ground support structure. Each leg of the system was designed to have 3 DOF for the hip, 1 DOF for the knee, and 3 DOF for the ankle. The degrees of freedom of the hips, knees, and ankles in the sagittal plane are linked to the gait motions and consist of active or quasi-passive joints within which power generation can be achieved. The other degrees of freedom are composed of passive joints for easy balance during gait motions. In addition, a concrete placing module is connected to the torso module. The concrete placing module is composed of the four-bar linkage, a spring, and holders that compensate the weight of the concrete placing hose. For the purpose of gait control, muscle circumference sensors (MCRSs) and insole sensors were developed for the HEXAR. The MCRSs measure the changes in the muscle's circumference during muscle contraction. The sensors can be worn on the clothes and customized for each worker. The insole sensor was developed for detecting the gait phase and measures the center of pressure (COP). The MCRS system is attached to the quadriceps muscle groups and hamstring muscle groups for measuring the intent signal for generating the commands for the joint control. The command signals for the HEXAR are generated by calculating the joint angles continuously.

System 135: Walking Assist Device with Bodyweight Support System

Developer: Honda Motor

Honda's robotic division has been since the 1980s working on advanced robotics technology and became famous through its many developments, experiments, and demonstrations in context with Honda's ASIMO bipedal, humanoid robot.

Figure 2.283. The walking assist device with bodyweight support system can support people in all kinds of daily activities including work activities. (Image: Honda Motor.)

Figure 2.284. Test of the walking assist device with bodyweight support system with workers in Honda's Saitama factory, in particular on the work-intensive final assembly line. (Image: Honda Motor.)

During the last decade, Honda used knowledge and technologies gained in the context of the ASIMO development (sensors, activators, control strategy, etc.) to develop a series of mobility assisting devices that can be considered wearable robots such as, for example, the bodyweight support assist device, the walking assist device, and the U3-X and UNI-CUB mobility devices. In particular, the bodyweight support assist device is intended and was demonstrated as being a device that can assist workers (Figures 2.283 and 2.284). The device consists of a central frame, two leg-links with shoes, a seat, and electrical motors. The user slips into the shoes and sits on the seat and the device then recognizes his movements and assists them. Furthermore, the device allows the user to go into certain positions and fix the system so that he can maintain this position assisted/supported by the device. Honda tested the device with workers in its Saitama factory, in particular, on the work-intensive final assembly line.

System 136: Supernumerary Robotic Limbs (SRLs)

Developer: d'Arabeloff Lab, MIT, PI: Professor H. Harry Asada

The supernumerary robotic limbs (SRLs) project follows a novel approach for wearable robots. Whereas with conventional exoskeletons the assistive robotic system more or less follows the movements of the limbs of the user, the SRL project develops wearable systems that provide additional limbs that operate independently from the arms and legs of the user. The SRLs are worn by the user as a backpack-like system. The backpack contains frames and profiles to which the limbs are attached and if necessary additional equipment such as battery packs. The images show that so far various prototypes have been realized and tested. One variant (Figure 2.285) provides strong limbs providing assistance from the hip area for different tasks (drilling, handling, etc.). A second variant (Figure 2.286) provides additional and highly dexterous limbs for assistance with overhead element installation. Because

Figure 2.285. SRL variant 1: Prototype used in a drilling task. (Image: © [2014] IEEE. Reprinted, with permission, from Parietti and Asada, 2014.)

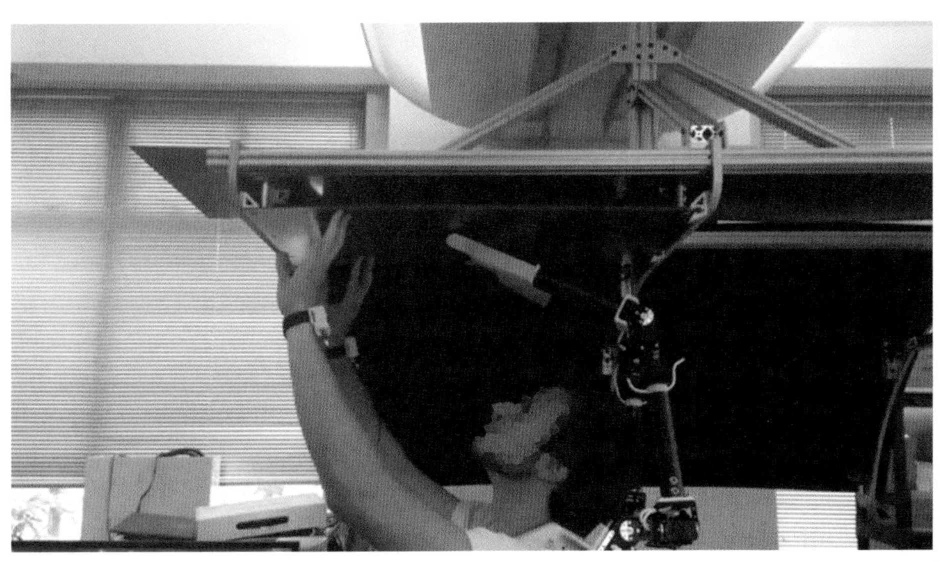

Figure 2.286. SRL variant 2 – highly dexterous limbs provide assistance with the overhead installation of interior elements. (Image: Baldin Llorens-Bonilla and Harry Asada, d'Arbeloff Lab.)

the limbs are not directly guided by human extremities, the developers explore novel modes of control such as, for example, predictive approaches allowing the system to recognize and predict human movements and work tasks and proactively bring the limbs into the right configuration and position. The limbs can be equipped with various kinds of end-effectors depending on the tasks to be accomplished. Currently, SRL research is directed at material handling tasks in manufacturing and in particular in the aircraft industry (Boeing). However, as outlined in **Volume 1**, the construction of aircraft and the construction of buildings have many things in common, and systems that can handle the installation of panels and interior parts in aircraft production can be used for similar tasks in the construction industry.

For further information, see Parietti and Asada (2014) and Llorens Bonilla and Asada (2014).

Systems 137–139: Smart Glasses

Developers: Various developers; for example, Vuzix, Google, Microsoft, Samsung, Oculus, and so forth

Leading technology companies are competing in developing smart glasses and see-through devices for industrial application. Besides use cases for offices, manufacturing, and logistics in general, use cases on the construction site are the focus of these companies and associated application developers. The tech start-up Bridgit (Bridgit 2015; Figure 2.290), for example, interfaced Google Glasses and Thalmic Labs' Myo armband to its construction site management software, allowing management of schedules, identification and logging of task/process specific issues on the site, and guiding workers and subcontractors through schedules and tasks. Vuzix has, with its M200AR waveguide HD device (Vuzix, 2015; Figures 2.287 and 2.288), developed a robust version of a see-through device that targets heavy equipment operators and

Figure 2.287. Vuzix M200AR waveguide HD device used by a utility installation worker at NS Solutions Corporation. (Image: NS Solutions Corporation.)

Figure 2.288. Vuzix M200AR waveguide HD device used by an electrician. (Image: Vuzix Corporation.)

assemblers and use for training purposes. The device comes with a software development kit that allows software specialists of the companies using the device to build their own, task-specific applications and augmented reality content. On the Microsoft Build 2015 (Autodesk, 2015; Figure 2.289), Autodesk, a major software company in

Figure 2.289. Microsoft HoloLens used in design and engineering. (Image: Microsoft.)

Figure 2.290. Bridgit (Bridgit 2015), for example, interfaced Google Glasses and Thalmic Labs' Myo armband to its construction site management software. (Image: Chair of Building Realisation and Robotics/C. Follini.)

the architecture and construction field, announced that it will develop the capability to load 3D models form its Autodesk software into the virtual reality environment of Microsoft HoloLens. Despite the fact that the current versions of smart glasses and see-through devices still lack usability (in terms of both direct control/use on site and software development), its use in the labour-based and flexibility demanding construction industry can be considered as an optimal solution for augmenting the capabilities of future knowledge workers in construction.

System 140: Affordable and Modular Cooperative Robots

Developer: Universal Robots

Universal Robots develops and manufactures affordable cooperative robots that can directly interact with human beings without the need for fences. Currently Universal Robots offers its UR robots in three variants: UR 3 (work radius: 0.50 m; payload: 3 kg), UR 5 (work radius: 0.85 m; payload: 5 kg), and UR 10 (work radius: 1.30 m; payload: 10 kg). In the general manufacturing industry the capabilities of the UR robots are currently explored and used by BMW in its factories and in particular in the final assembly areas (Figure 2.291). Also in the context of on-site construction several institutions (e.g., ETH Zürich; see Section 2.17) and companies (e.g., nLink, see Section 2.21) utilize the capabilities of the UR robots. What makes the robots interesting for on-site construction are their easy and quick programmability (e.g., through teaching), its collaborative features, and its simplicity and affordability.

Figure 2.291. UR robots assist human workers with assembly tasks in factories of the BMW Group. (Image: BMW Group.)

2.21 Interior Finishing Robots

Interior finishing work is, both in terms of ergonomics and of productivity, unfavourable work, as cranes or other machines used to lift and position parts and components can rarely be used. The first series of interior finishing robots was deployed between 1988 and 1994, and the development continued in parallel with the development and deployment of integrated automated sites, especially as interior finishing robots were used within these automated construction sites for finishing tasks (see also **Volume 4**). According to Cousineau and Miura (1998), the first series of ceiling panel installation robots deployed in the 1980s and 1990s achieved only marginal improvements in terms of speed (e.g., CFR-1: reduced time required for installing a panel from 3–4 minutes down to about 2.5 minutes) and human labour requirements (e.g., CFR-1: only one human worker necessary instead of two). However, research and development continued and the latest approaches (e.g., nLink's Mobile Drilling Robot) show that advances in robotics as well as in the integration of STCRs with work planning software is finally able to produce robotic system that are feasible and cost-effective alternatives to conventional construction. Systems in this category include a broad variety of systems such as mobile platforms equipped with a manipulator that are used to handle and position wall panels; robots that assist with or are able to fully automate the installation of ceiling elements; robots that assist with the installation of large plumbing or aeration system elements; systems for the application of wall papers, sheets, or membranes; systems for the application of mortar/plaster to walls; and robots for the automated drilling of holes into walls and ceilings. Owing to the large variety of tasks associated with interior finishing, some of the systems were even designed as modular kits that can be customized to or adapted quickly to various site conditions and tasks.

Interior Finishing Robots		
141	Ceiling handling robot	Hazama
142	Mobile plasterboard handling robot	Komatsu
143	Mini crane robot, Kalcatta, LM15-1	Komatsu
144–147	Multipurpose construction robots (LH Series) • LH 30 • LH 50 • LH 80 • LH 150	Kajima and Komatsu
148	Lightweight manipulator for ceiling and wall panel installation	Tokyu Construction Co., Ltd.
149	Robot for automated ceiling panel installation	Tokyu Construction Co., Ltd.
150	Ceiling panel installation robot	Shimizu
151	Plumbing part positioning robot	Komatsu
152	Interior panel handling robot	Taisei
153	Wallpapering robot	Komatsu
154	Mobile robot for installation of heavy components	Komatsu
155	Mobile interior finishing robot	Lindner/TUM
156	Mobile material handling robot Boardman	Taisei
157	Automatic rendering machine	Anex Industrial Hong Kong Limited and EZ Renda Construction Machinery Limited
158–159	Robots for the automated installation of partition-wall framing (Trackbot and Studmaster)	MIT
160	Mobile drilling robot	nLink AS

System 141: Ceiling Handling Robot

Developer: Hazama Ando Corporation

Holding ceiling panels, for example, plaster boards, in the correct position while standing on an elevated platform to perform positioning, adjusting, and fixation work is time and labour intensive and ergonomically challenging. Often this task requires multiple workers. The ceiling handling robot (Figure 2.292) by Hazama addresses this and simplifies the entire process. Ceiling panel boards are loaded (e.g., by a fork-lift) into a "board storage platform" and then lifted up to the required height. Each ceiling board then is automatically shifted horizontally over to the end-effector. The end-effector is directly guided/controlled by a human worker and allows positioning, adjusting, and fixing (through integrated multiple nailing guns) of the panels. The whole system is based on a mobile robotic platform that can be controlled and moved by the human worker through a control panel allocated near the end-effector. Both the board storage platform and the board end-effector are height adjustable and can thus be adjusted to a variety of ceiling heights.

System 142: Mobile Plasterboard Handling Robot

Developer: Komatsu Ltd.

The robot system (Figure 2.293) was developed by Komatsu for assisting construction workers or craftsmen with the installation of plasterboards to walls. The

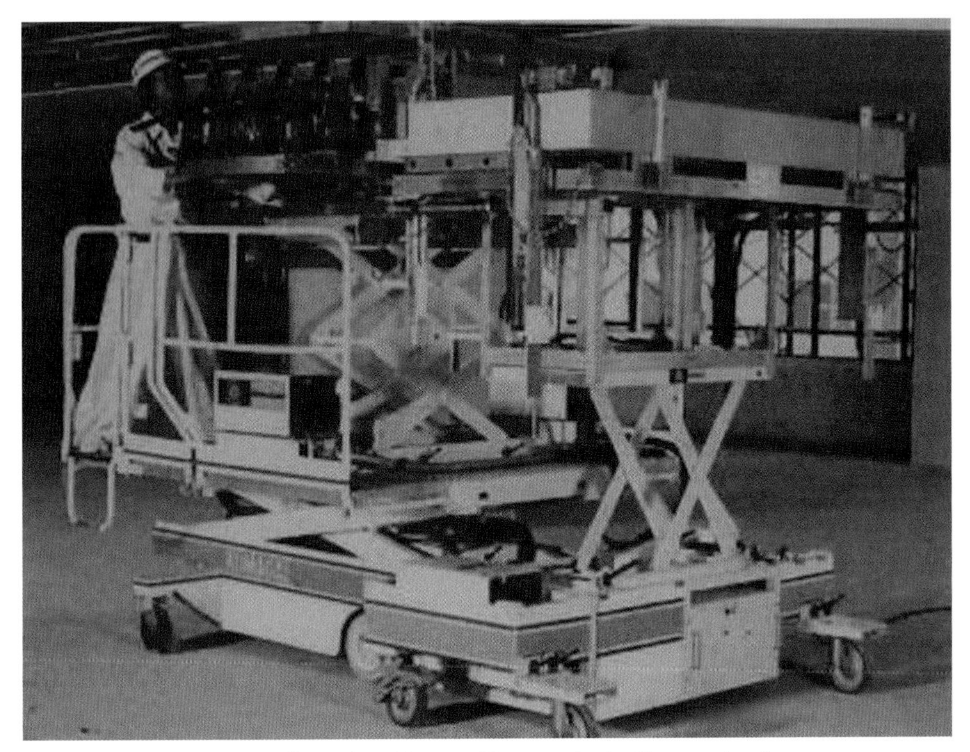

Figure 2.292. Ceiling handling robot. (Image: Hazama Ando Corporation.)

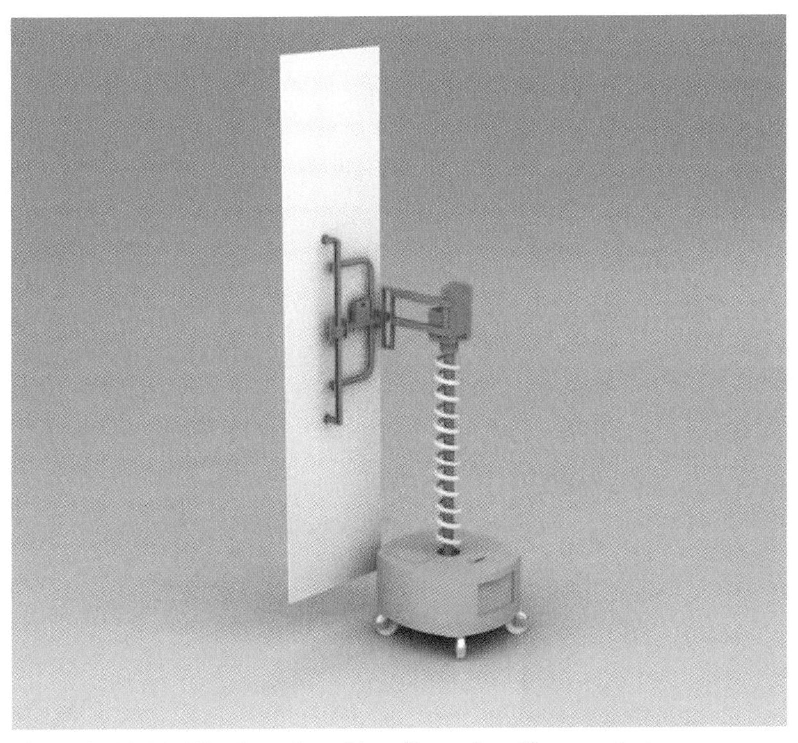

Figure 2.293. Mobile plasterboard handling robot, Komatsu Ltd.

Figure 2.294. Mini crane robot Kalcatta, LM15-1.

system is highly compact and lightweight and consists of a mobile base (without activators), a manipulator, and an end-effector with suction cups to pick up, hold, and release the panels. Owing to its compactness and light weight the robot can be moved by the worker though the work area manually and the manipulator and end-effector can be brought by the worker through direct force control into the desired positions/orientations. The system allows one to position and adjust the panel in question and hold the steel panel in the desired position during the manual fixation process.

System 143: Mini Crane Robot Kalcatta, LM15-1

Developer: Komatsu Ltd.

This small-sized assistive robotic crane (dimensions: L 0.96 m, W 0.79 m, H 1.97 m; Figure 2.294) was developed to assist construction workers in narrow spaces and particularly to install concrete panels. Its application in the industry lowers manpower requirements and allows a single operator to handle relatively heavy panels. The robot can be folded into a compact shape for transportation, and can thus be transported inside buildings or by a van. The system is modular and can be disassembled into parts if required for transportation (individual parts of the robot have the following weights: crane unit: 300 kg, crawler unit: 150 kg, outrigger: 70 kg). The robot can be operated by wireless remote control, and it can be used to lift components of a weight of up to 350 kilograms.

For further information, see Council for Construction Robot Research (1999, pp. 36–37).

Systems 144–147: Multipurpose Construction Robots (LH Series)

Developer: Kajima Corporation and Komatsu Ltd.

The contractor Kajima and the machine and equipment producer Komatsu jointly developed the LH handling robot series. The LH series (LH 30, LH 50, LH 80, and LH 150; Figures 2.295–2.300) covers all payloads related to the handling of

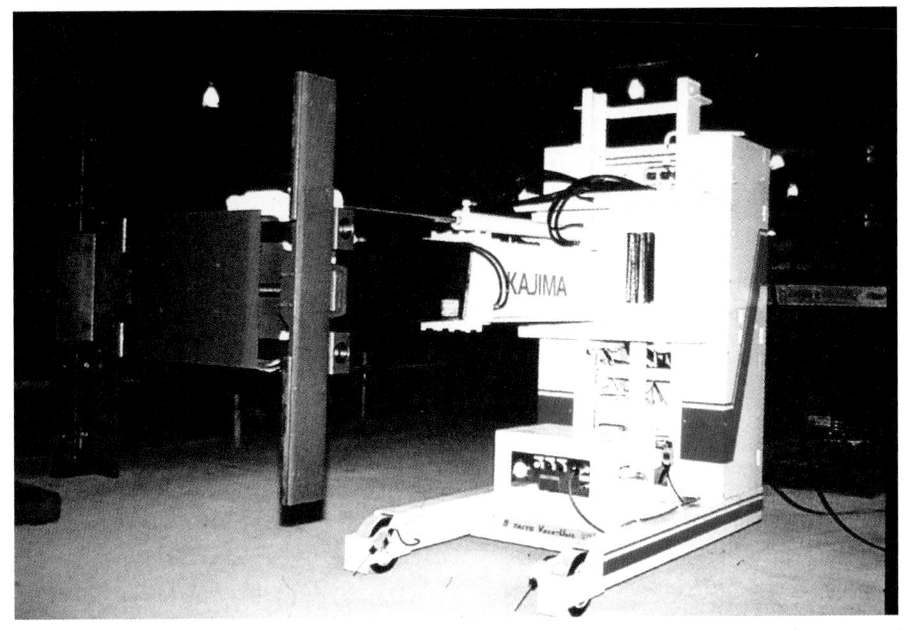

Figure 2.295. Multipurpose robot equipped with panel handling end-effector. (Image: Kajima Corporation.)

Figure 2.296. Detailed view of the end-effector for multipurpose construction robots.

Figure 2.297: Robot for setting concrete columns (LH 50). (Image: Kajima Corporation.)

Figure 2.298. Interior finishing material handling system (LH 80), back view.

Figure 2.299. Robot for setting concrete walls (LH 150). (Image: Kajima Corporation.)

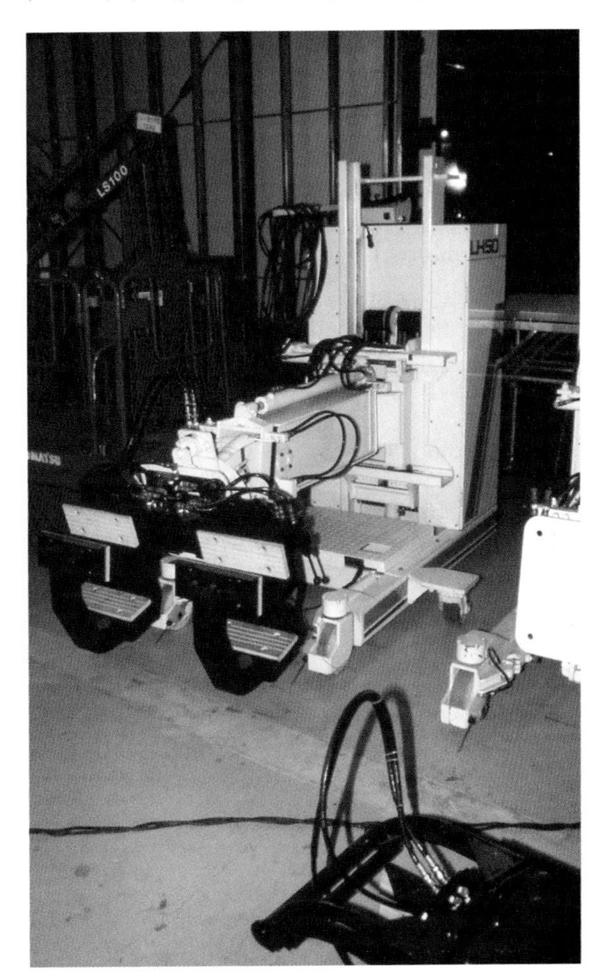

Figure 2.300. Robot for setting concrete walls (LH 150), an end–effector.

Figure 2.301. Lightweight manipulator used to install ceiling panels. (Image: Tokyu Construction Co., Ltd.)

smaller and larger parts, components, and panels within individual floor levels. The basic idea for all of the robots is the same. A mobile base frame is equipped with a manipulator that can travel (translational movement) vertically along the mobile base frame. A variety of end-effectors can be attached and detached to the manipulators within each LH payload category. By combining different LH payload categories and the modular collection of end-effectors, a large set of use cases and processes can be served.

System 148: Lightweight Manipulator for Ceiling and Wall Panel Installation

Developer: Tokyu Construction Co., Ltd.

This robot system was developed to assist workers with installing ceiling (Figure 2.301) and wall (Figure 2.302) panels during the interior finishing process. The system is not fully autonomous but assists the worker so that one single worker can efficiently install ceiling panels without assistance from other workers. The robot system (its movement on the floor as well as the positioning system) is guided by the worker, thus combining the power of the robot system with the flexibility and dexterity of the human worker. The system consists of a mobile robotic base, a manipulator, a movable platform connected to the mobile robotic platform, an end-effector equipped with vacuum suction components, and a handling interface installed on the end-effector. During the panel installation process, a human worker stands safely on the movable platform and guides both the manipulator and the end-effector by

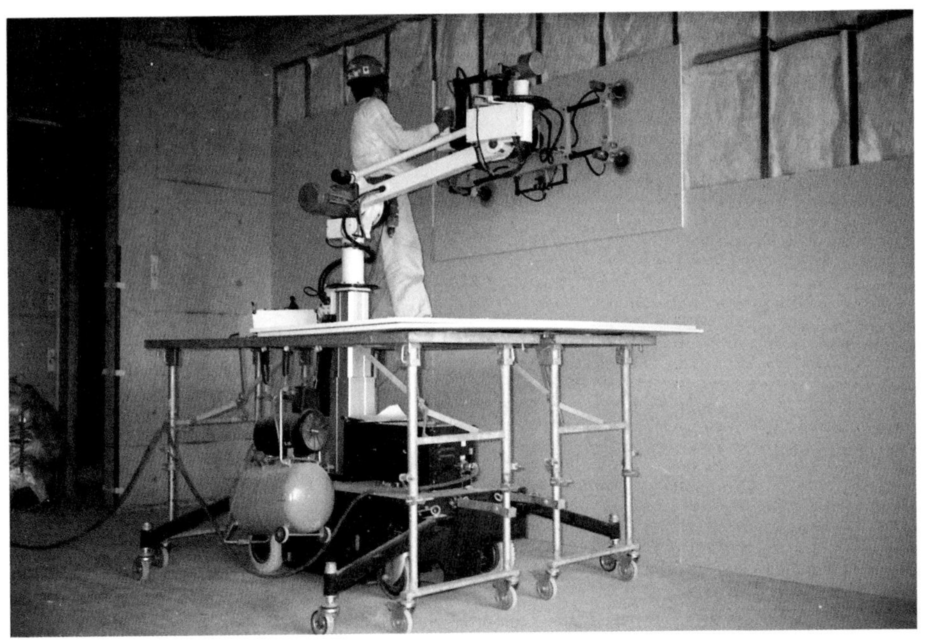

Figure 2.302. Tokyu lightweight manipulator used to install wall panels. (Tokyu Construction Co., Ltd.)

means of the handling interface. Both the movable platform and the manipulator can be adjusted to various wall/ceiling heights.

For further information, see Council for Construction Robot Research (1999, pp. 260–261).

System 149: Robot for Fully Automated Ceiling Panel Installation

Developer: Tokyu Construction Co., Ltd.

Tokyu Construction developed this robot for the full automation of the process of ceiling panel installation (Figure 2.303). The robot in that context automates micro-logistics (temporary storage and the presentation of the panels to the installation system), the positioning of the panels, as well as their fixation through bolting. The robot consists of a mobile platform able to conduct autonomous repositioning of the system along the work trajectory, a magazine for temporary storage of ceiling panels that makes it possible to present them to the end-effector, a manipulator that is able to position and adjust the panels, and an end-effector that allows pick up, holding, and fixing of the panels. Electric power comes from an external source through a cable roller. The end-effector contains small wheels that allow the panels to be shifted through a mechanism installed to the storage magazine over to the end-effector. Once the panel rests on the end-effector it is held by vacuum suction cups. Through cooperative motion of the mobile base and the manipulator the end-effector is bought into position before the end-effector's bolting mechanism fixes the panels to the ceiling's substructure.

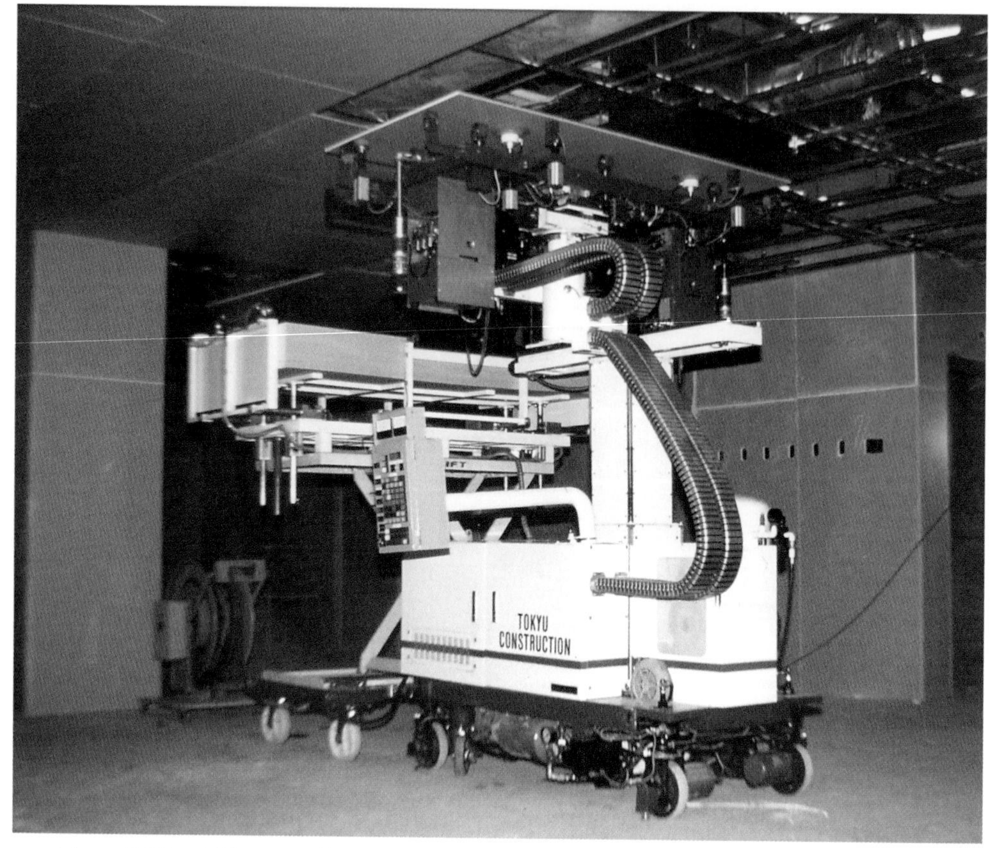

Figure 2.303. Ceiling panel installation robot. (Image: Tokyu Construction Co., Ltd.)

System 150: Ceiling Panel Positioning Robot CFR-1

Developer: Shimizu Corporation

The ceiling panel installation robot developed by Shimizu is used to assist in the installation of ceiling panels (Figures 2.304–2.306). The system is compact and lightweight (total weight without panels: about 350 kg) and can be easily moved on the site. The system consists of a robotic mobile platform (dimensions: L 1.10 m × W 0.60 m), a main body (dimensions: L 1.38 m × W 0.70 mm× H 1.75 m), a panel handling mechanism (lifting arm, dimensions: L 1.90 m × W 0.50 m × H 1.00m, lead capacity: up to 300 kg), an end-effector (panel holder, equipped with a compliant mechanism that during panel positioning allows adjustment of the panel position to the position of surrounding panels/edges), an on-board panel storage magazine, and a remote control unit. The panel-handling device consists of a lifting arm and an end-effector for grasping the panels. The robot can carry up to 20 boards at one time in its on-board panel storage magazine. The robot grasps, raises, and positions/adjusts the boards automatically, initiated remotely through a control panel. Once positioned and adjusted by the robot the boards must be fixed to the ceiling manually by a worker. Through the robotic mobile platform the robot can be automatically repositioned according to predefined panel dimensions. The system is able to serve ceiling heights from 2.4 to 3.5 metres and allows installation of about 25 panels/hour.

For further information, see Ueno et al. (1988).

Figure 2.304. Picking of element from magazine by CFR-1.

Figure 2.305. Lifting of element by CFR-1.

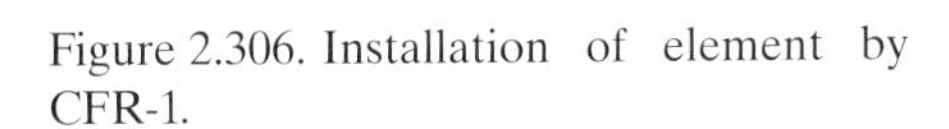

Figure 2.306. Installation of element by CFR-1.

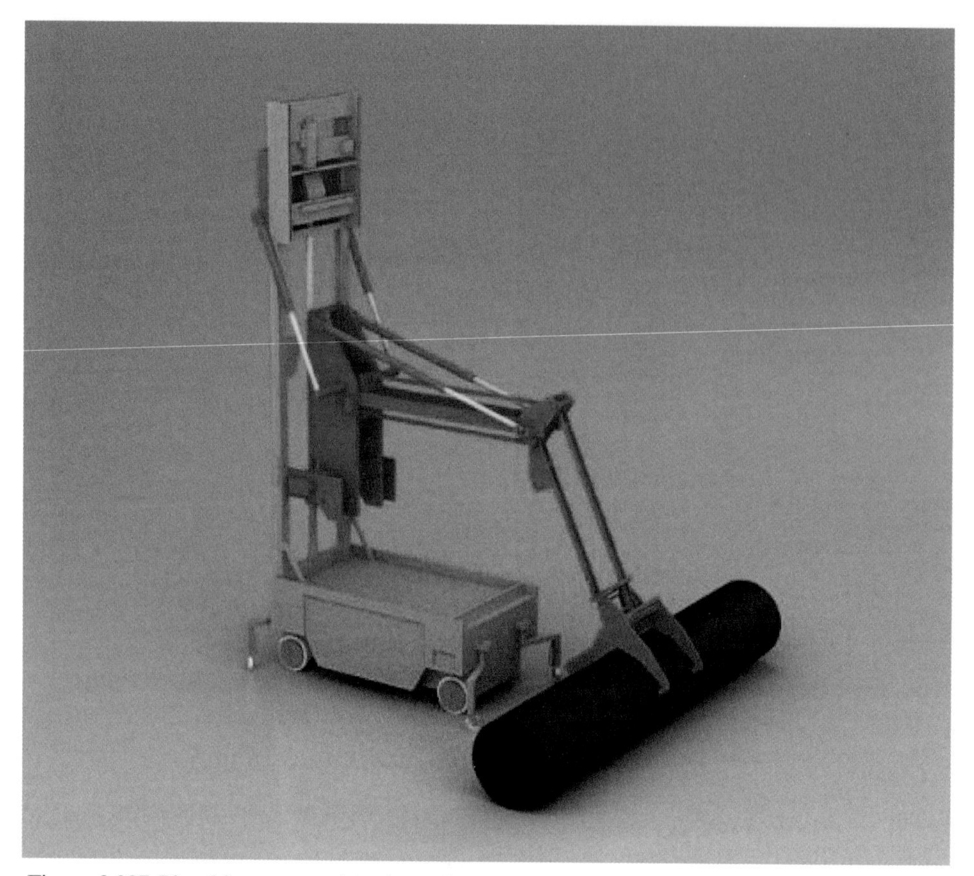

Figure 2.307. Plumbing part positioning robot, Komatsu Ltd.

System 151: Plumbing Part Positioning Robot

Developer: Komatsu Ltd.

The Komatsu plumbing part positioning robot (Figure 2.307) is designed to reduce the amount of human labour needed on the construction site and to speed up the process of the installation of larger plumbing elements. Usually a team of several workers cooperatively has to perform such plumbing tasks, including preliminary installation of a scaffolding system and the positioning and adjustment of the plumbing elements. This robot system allows one single worker to carry out the installation of plumbing elements. It consists of a mobile platform, a frame on top of the mobile platform, a multi-DOF manipulator that can be lifted up and down along the platform, and an end-effector for grasping plumbing elements. A single worker manually operates the robot via remote control. The robot is able to position the plumbing elements and holds them in the desired position while the worker fixes the element.

System 152: Interior Panel Handling Robot

Developer: Taisei Corporation

Taisei developed this interior panel-handling robot (Figure 2.308) to assist workers with the installation of plasterboard elements on walls in a human–robot

Figure 2.308. 3D sketch of interior panel handling robot.

cooperative manner. A worker guides the robot and its end-effector. The worker only needs to exert minimal force to guide the robot system. Once the position/orientation of the plasterboard element is fine tuned, the robot system can hold the plasterboard in this intended final position during the fixation process. The robot system consists of a robotic mobile platform with extendable outriggers, a manipulator with 5 DOF, an end-effector with suction components for picking up and releasing the plasterboard panels, and a control device installed on the end-effector from which the robot and end-effector can be controlled/guided.

System 153: Wallpapering Robot

Developer: Komatsu Ltd.

Wallpapering is a task that requires skilled labour to be performed with a high quality. To achieve high-quality results while simultaneously reducing financial spending on hiring highly skilled labourers, Komatsu developed a robot for automating the wallpapering process (Figure 2.309). The robot moves alongside the wall on a mobile platform to install the wallpaper automatically, vertical section by vertical section. The robot aligns itself to the targeted wall area with the help of a distance sensor system that is able to scan the surroundings to identify the working area. A rollers-based system at the front end of the robot helps to maintain a fixed distance of the system to the wall. The robot is supplied from its lower back section with wallpaper (from a roll). A sheet of wallpaper taken from the roll by the robot and glue are applied by the system. Depending on the size of the wall an automatic cutter separates the sheet from the rest of the roll, and the installation end-effector slides vertically to fix the wallpaper sheet.

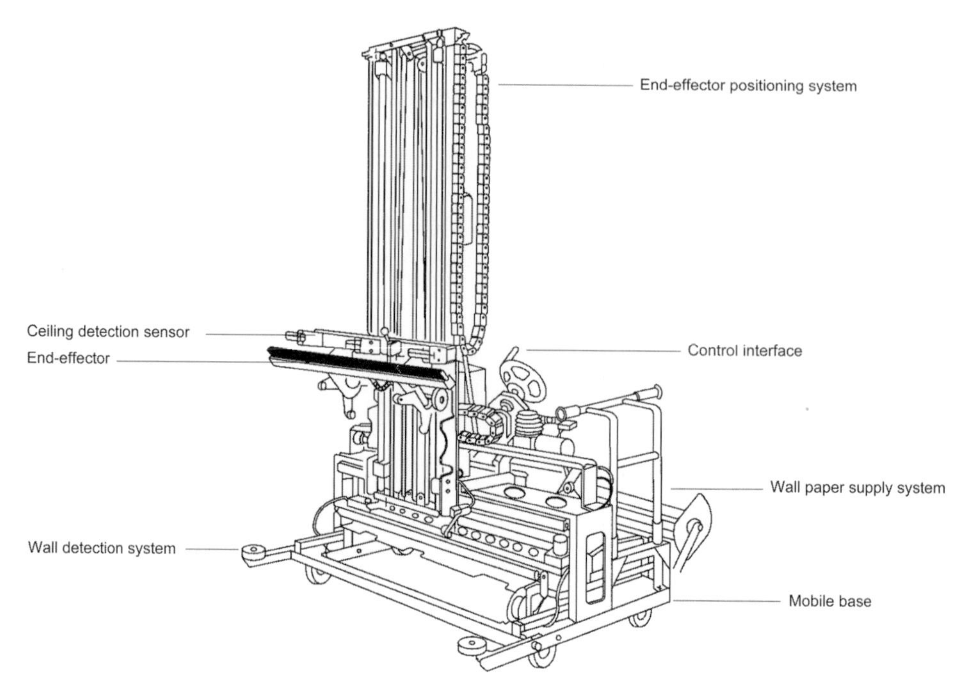

Figure 2.309. Wallpapering robot, Komatsu Ltd. (Image: Chair of Building Realization and Robotics. Redrawn based on Hasegawa 1999.)

For further information, see Hasegawa (1999).

System 154: Mobile Robot for Installation of Heavy Components

Developer: Komatsu Ltd.

Komatsu developed this robot system for the handling of relatively large and heavy reinforced concrete elements (Figure 2.310). It can be used for building

Figure 2.310. Mobile robot for installation of heavy components, Komatsu Ltd.

Figure 2.311. Overview of mobile and modular interior finishing robot kit system.

construction projects as well as within the infrastructure construction domain. The robot system consists of a tracked mobile robotic platform, a hydraulic manipulator with 3 DOF, and an end-effector for grasping various types of concrete components. The manipulator is able to position the component, and using an additional degree of freedom integrated into the end-effector, orient the component to the left or right. The end-effector is able pick up and balance a variety of different sizes of components. One requirement for this robot system is that components can be picked up and installed from below.

System 155: Mobile and Modular Interior Finishing Robot Kit

Developer: Lindner/TUM

The system (Figures 2.311 and 2.312) was designed in a modular robot kit that allows a robotic base to be equipped with a variety of sensors and working modules/end-effectors and thus to be customized to a large variety of interior

Figure 2.312. Base system equipped with a working module for overhead nailing.

finishing tasks and individual site conditions. The system consists of a mobile platform, a frame that allows lowering/lifting and positioning of the working module, and an exchangeable working module frame. Working module frames can be equipped with a variety of tools and end-effectors that allow the automation of tasks such as ceiling/wall panel installation, nailing, and cleaning. The working modules can be customized to certain site conditions and the change of working modules can be done within a few minutes so that the working mode of the robot can be changed on the site. A tracked mobile platform controls the movement and orientation of the robot to the given environment and contains modular compartments that can, depending on the task to be performed by the robot, be equipped with sensors such as laser-based detection sensors, infrared, and ultrasonic systems.

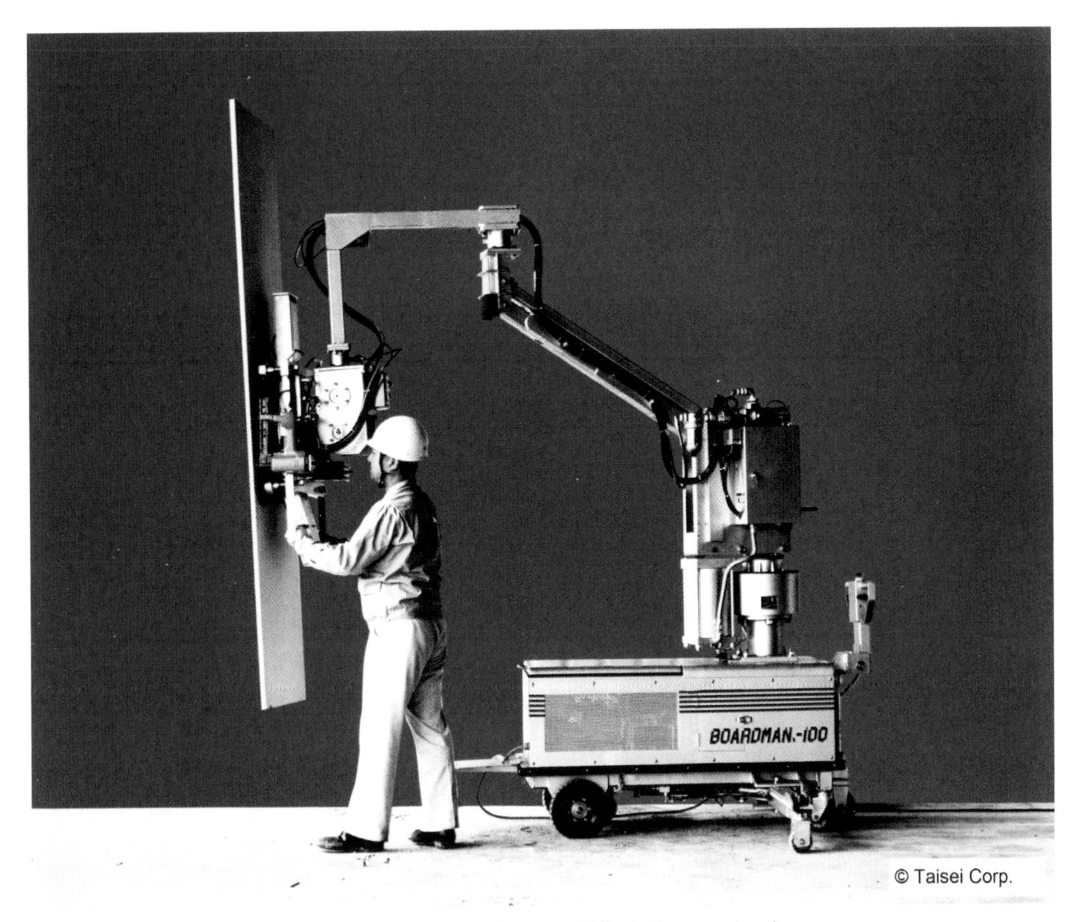

Figure 2.313. Mobile material handling robot. (Image: Taisei Corporation.)

System 156: Mobile Material Handling Robot Boardman

Developer: Taisei Corporation

Taisei's mobile material handling robot is used for the handling of relatively lightweight panels in the exterior finishing process (Figure 2.313). The robot system can be used to pick up, transport, position, and align those panels. It consists of a mobile platform, a multijoint manipulator, and an end-effector. A correctly positioned and aligned panel can be held safely in its final position by the robot system until the worker operating the system can affix the panel. The robot system can reduce the number of workers necessary to conduct the installation operation. The system also relieves the workers from physically demanding work and allows the operator to concentrate on the installation process. The end-effector grips and releases the handled panels by a set of vacuum suction devices. The end-effector also contains the handling interface that allows the operator, with minimal force, to guide the manipulator and the end-effector into place and to position and align the handled panels (Figure 2.314).

Figure 2.314. Mobile material handling robot. (Image: Taisei Corporation.)

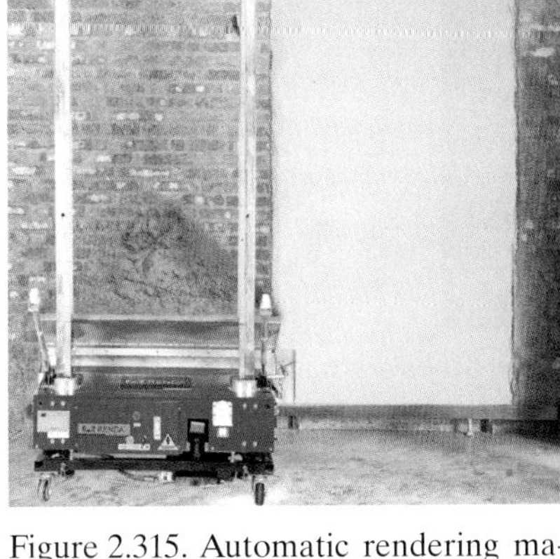

Figure 2.315. Automatic rendering machine in action. (Anex Industrial Hong Kong Limited, manufacturer: EZ Renda Construction Machinery Limited.)

Figure 2.316. Automatic rendering machine. (Anex Industrial Hong Kong Limited, manufacturer: EZ Renda Construction Machinery Limited.)

System 157: Automatic Rendering Machine

Developer: Anex Industrial Hong Kong Limited and EZ Renda Construction Machinery Limited

Automatic rendering machines were developed and manufactured by a Hong Kong–based company to support and partly automate the labour-intensive task of wall plastering (Figures 2.315 and 2.316). The machine is compact and can be handled and repositioned by one operator. Positioning/repositioning is done manually. The machine is fixed between floor and ceiling and then along two guiding rails moves the end-effector slowly upwards to distribute and fix the plaster to the wall. If demanded the plastered wall segments can be reworked manually while the machine is placing the plaster at a subsequent segment. The machine is available in different sizes and versions to suit individual site conditions.

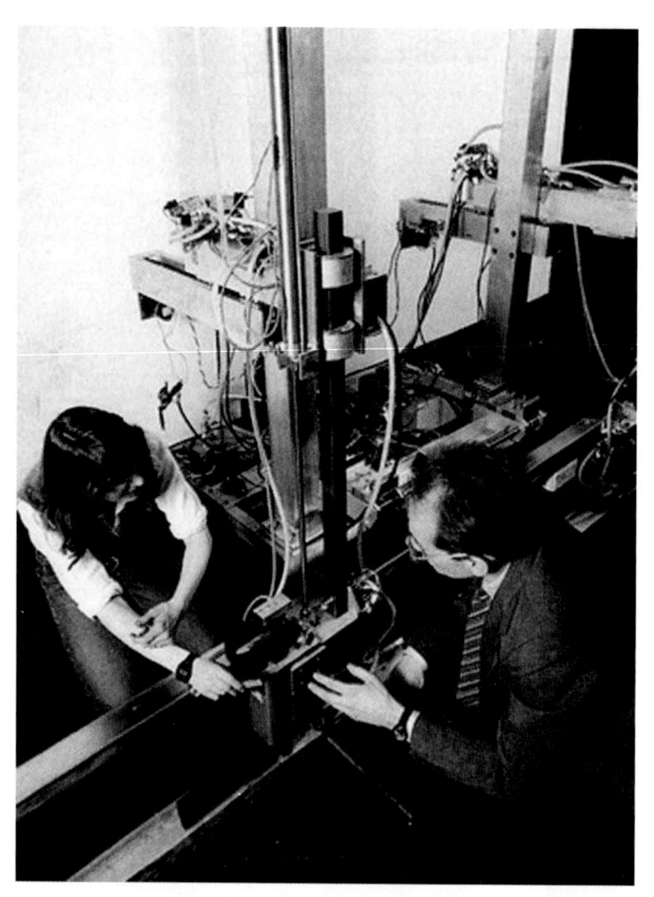

Figure 2.317. The Trackbot places and fixes studs to walls to floors and ceilings. (Image: originally published by Brown, S. F. (1988) "Meet the Wallbots", in: Popular Science, August 1988, pp. 54–55. Photographer: Greg Sharko.)

Systems 158 and 159: Robots for the Automated Installation of Partition-Wall Framing (Trackbot and Studbot)

Developer: MIT (Laura Demsetz, Andy Ziegler, David Levy, Bruce Schena, Alex Slocum)

The system was designed to fully automate the installation of partition-wall framing in large buildings such as office buildings or hospitals. The system consists of two robots, one specialized in the installation of c-shaped tracks to the floor and ceiling (Trackbot) and one robot specialized in the installation of the studs between the tracks on the floor and the ceiling that the Trackbot previously installed (Studbot).

The Trackbot (Figure 2.317) consists of a mobile base, on-board storage for about 10 tracks, and a frame along which the placement/fixation end-effectors (a set of two end-effectors for the floor and a set of two end-effectors for the ceiling) can be moved. A vacuum gripper retrieves the tracks from the storage and places them on the ground before a pair of nailing guns fixes them to the floor. The robot was equipped with photodetection sensors that allow it to follow and operate

Figure 2.318. The Studbot installs the studs between the tracks the Trackbot previously installed to floors and ceilings. (Image: Originally published by Brown, S. F. (1988) "Meet the Wallbots", in: Popular Science, August 1988, pp. 54–55. Photographer: Greg Sharko.)

along a marking created on the floor and the ceiling by an externally set up rotating laser.

The Studbot (Figure 2.318) consists of a mobile base, on-board storage for about 100 studs, and a manipulator equipped with an end-effector to grasp and position the studs in a vertical position between the previously installed tracks. The manipulator rotates the studs from a horizontal position into the desired vertical position in a way that they are positioned correctly between the c-shaped floor and ceiling tracks. The Studbot is guided by the previously installed tracks, and the locations where studs are to be placed are preprogrammed according to the installation plans.

For further information, see Slocum et al. (1987) and Brown (1988).

System 160: Mobile Drilling Robot

Developer: nLink AS

The mobile drilling robot from nLink AS automated the process of measuring, marking, and drilling holes into ceilings for the fixation of light elements, ventilation,

Figure 2.319. Newest version of nLink's mobile drilling robot equipped with a mobile robotic platform. (Image: Håvard Halvorsen, © nLink AS.)

ceiling panels, and other types of equipment (Figures 2.319 and 2.320). In the construction of large buildings the number of holes to be drilled into ceilings is enormous, and the work is time consuming. From an ergonomic perspective this is not optimal for being conducted by human beings. The robot system is laser guided by both an internal and external laser system and can be placed on a variety of mobile electric/robotic or manual platforms (e.g., scissor lifts). The user interface on the tablet allows the operator to select certain drilling patterns, or to preprogram the drilling path from BIM files. The robot is repositioned by a human worker who only needs to do rough positioning, and the robot finds the exact drilling position in an automated manner. The end-effector can be equipped with various types of power tools that are combined with a device that during drilling vacuums dust and powder from the drilling process.

Figure 2.320. nLink's mobile drilling robot in operation. (Image: Håvard Halvorsen, © nLink AS.)

2.22 Fireproof Coating Robots

In many countries, building regulations demand that steel structures are covered with fireproof plates and/or fire prevention paint. These measures can be taken only after the steel parts and elements have been joined on site to be able to connect elements properly (e.g., by welding) and avoid any damage to the fireproof coating during assembly operations. A shift of fireproof coating operations off-site production steps, as in a structured factory environment where coating could be completed with high efficiency, is not practical. Particularly in countries where earthquakes and a high rate of high-rise construction encourage the extensive use of steel structures, the development and employment of automated and robotic systems able to coat steel structures, after they have been erected on site, are a feasible solution. Development of systems such as the SSR1, SSR2, and the SSR3 in this area of robots continued from the 1980s up to today. Systems in this category can be subclassified into two major subcategories. Systems in the one subcategory build on robotic manipulators mounted on top of mobile platforms that can follow the elements to be coated. Systems in the other category are attached directly to the beam or column and thus move along the element they are coating. A common element of all systems is that they require an on-site deployable system that pumps the coating material to the robot.

Fireproof Coating Robots		
161	Shimizu fireproof coating robots Shimizu Spray-painting Robot (SSR1, SSR2, SSR3)	Shimizu
162	Fireproof coating and rock-wool spraying mobile robot	Fujita
163	TN-Hukkun fireproof coating robot	Toda
164	Fireproof and rock-wool spraying beam-attached/rail-guided robot	Fujita

System 161: Shimizu Fireproof Coating Robots (SSR1, SSR2, SSR3)

Developer: Shimizu Corporation

Shimizu developed the SSR robot series (Figures 2.321–2.324) to automate and speed up the fireproof coating of steel structures, in particular, within the building finishing process. The SSR1 was the first version developed at the beginning of the 1980s and the SSR3 was the latest version developed at the end of the 1990s. Because Shimizu is one of the largest contractors in Japan, and involved in the construction of

Figure 2.321. Shimizu fireproof coating robots in action.

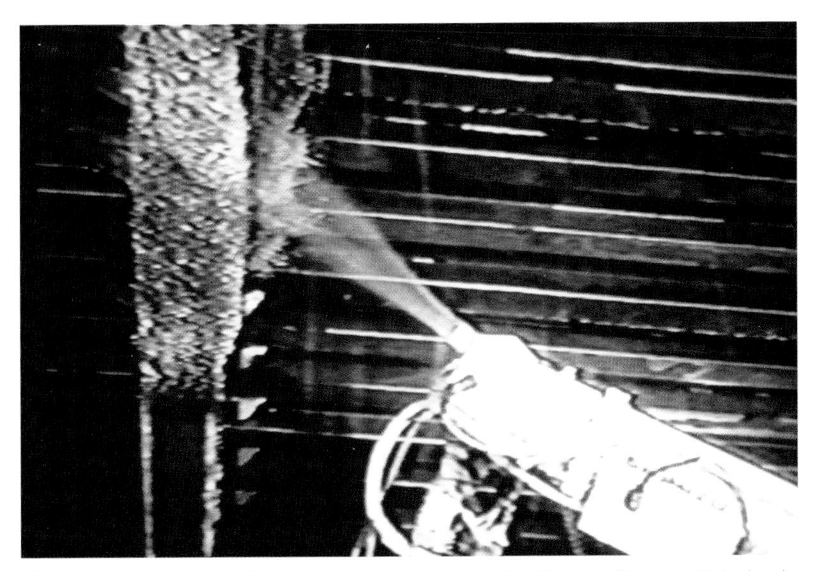

Figure 2.322. Shimizu fireproof coating robot end-effector. (Image: Shimizu.)

Figure 2.323. Side view of Shimizu fireproof coating robot system.

Figure 2.324. System for the supply of coating material.

a multitude of steel-based high-rise buildings, the automation of the fireproof coating process is key for improving Shimizu's efficiency and performance in those projects. The robot system (dimensions: L 1.60 m × W 1.05 m × H 2.145 m; weight: 1000 kg) consists of a mobile robotic platform (2 DOF, DC servo motor, travelling speed: max. 20 m/minute, position accuracy: ±5 mm), a lifting platform for the manipulator, the manipulator itself (4 DOF, DC servo motors, repeatability: ±3 mm), an end-effector for spraying the rock wool coating, a pole equipped with sensors mounted to the mobile platform, and a stationary plant module that pumps the rock wool coating through a hose system to the robot and its end-effector. The lifting platform allows the system to be adjusted for a variety of room heights. The manipulator provides the system with enough dexterity to work on a variety of beam and column shapes. The pole equipped with sensors allows the system to detect and autonomously follow, for example, the beams that it must spray. The stationary plant consists of a mixing unit and pumping unit and can be temporarily positioned on the floor level where the system operates.

For further information, see Council for Construction Robot Research (1999, pp. 240–241).

System 162: Fireproof Coating and Rock-Wool Spraying Mobile Robot

Developer: Fujita Corporation

This system was developed to decrease labour requirements and improve the quality of fireproofing coatings. The system consists of a mobile platform (dimensions: L 4.727 m × W 0.80 m × H 1.80 m), a 9-DOF manipulator equipped with a spraying end-effector nozzle, a laser-based component detection system, and a control interface (installed on the rear end of the mobile platform) for data input and

Figure 2.325. Fireproofing insulation spray robot. (Image: Fujita Corporation/Council for Construction Robot Research, 1999, pp. 218–219.)

adjustment of site-specific parameters (Figure 2.325). The manipulator consists of a lifting system for rough positioning or a 7-DOF robotic arm that can be operated with a speed of up to 1.55 metres per second. The laser-based component detection system is installed on two poles on the spraying side's end points of the mobile platform and helps to identify and detect the position of the steel elements to be coated and to guide the robot along these elements. A single worker can operate the robot and conduct the required data input through the control interface panel. The robot process steel elements in a work area of up to a 4-metre height if the work area is free of obstacles.

For further information, see Council for Construction Robot Research (1999, pp. 218–219).

System 163: TN-Fukkun Fireproof Coating Robot

Developer: Toda Corporation

The fireproof coating robot (Figure 2.326) from Toda is used to spray fire-resistant rock wool on steel beams as a fireproof coating. The system consists of a robotic mobile platform, a robotic manipulator with a spraying end-effector for applying the coating, and separate stationary coating supply unit that could if necessary also supply more than one robot. The mobile platform allows for the mobility of the system. The robotic manipulator has 6 DOF and thus provides the the ability to process also steel components with a more complex geometry. The manipulator is equipped with a spray gun end-effector to apply the fireproof coating. The system is programmed and operated from a separate, stationary control station.

For further information, see Council for Construction Robot Research (1999, pp. 258–259).

Figure 2.326. TN-Fukkun Fireproofing robot. (Image: Toda Corporation.)

System164: Fireproof and Rock-Wool Spraying Beam-Attached/Rail-Guided Robot

Developer: Fujita Corporation

The system is compact and can travel along the beams to be coated (Figure 2.327). The robot consists of a travel mechanism, a manipulator equipped with a spray

Figure 2.327. Fireproof and rock-wool spraying beam-attached/rail-guided robot. (Image: Fujita Corporation.)

gun end-effector, a laser-based automatic guidance system, and a separate stationary control desk. The travel mechanism travels along a beam/girder and thus moves and positions the manipulator. The laser-based automatic guidance system allows adjustment of the parameter related to operation speed and spray distance. In contrast to other robot systems in this category this robot system should be used before earlier construction phases, directly after the erection of the steel structure.

2.23 Service, Maintenance, and Inspection Robots

Facades of high-rise buildings are often clad with tiles, glass panels, or other surfaces that must be inspected, serviced, and maintained regularly throughout the building's life cycle. In particular, inspections are necessary to detect structural damage and to exchange tiles or panels that are at risk of falling to the ground. Also, with the increase of glass surfaces and heat input, facades are supposed to provide an important functional element that must be cleaned and maintained regularly. Typically, workers access facades via cages or gondolas suspended from the roof and the work is usually considered monotonous, inefficient, and dangerous and therefore often allocated to the low-wage work sector. Service, maintenance, and inspection robots are able to execute these monotonous and dangerous tasks autonomously. In many instances, in particular in the context of inspection, these systems also proved more reliable and provide a large amount of detailed data. For the inspection of a 40-metres high building facade (3000 m^2 surface area) a robot required an average of 8 hours, including approximately 1 hour for preparation, configuration, conversions, dismantling, and cleaning of the robot (Bock et al. 2010). According to a study conducted by the Architectural Institute of Japan (AIJ 1990; also addressed by Cousineau & Miura 1998), the major weaknesses of facade inspection robots were the large amount of human labour and set-up time required to install, program, calibrate, supervise, and uninstall the systems within an unstructured construction environment. Robots in this category cover a broad array of systems ranging from facade cleaning and inspection robots (cable suspended, facade climbing, rail guided, etc.), to rescue and firefighting solutions, furniture rearrangement robots, and robots for the inspection of the aeration system in the building.

System 165: Tile Facade Inspection Robot

Developer: Obayashi Corporation

The tile facade inspection robot from Obayashi (dimensions: L 0.725 m × W 0.665 m × H 0.265 m ; Figure 2.328) is used to automatically inspect the condition of each individual tile within a tiled exterior facade. It can be used to identify the level of deterioration of a building's facade and proactively prepare an adequate repair schedule. The system consists of the main robot unit, a system located at the roof top for suspending/positioning the main robot unit. The main robot unit is fixed temporarily to the wall during actual inspection by a vacuum suction system and is equipped with system for sound-based tile inspection consisting of a striking hammer as well as a microphone system. The detected sound reflection is used to analyse the condition of the tiles. The robot is also equipped with distance sensors to detect obstacles in

Service, Maintenance, and Inspection Robots		
165	Tile facade inspection robot	Obayashi
166	Facade inspection robot	Kajima
167	Facade inspection robot	Taisei
168	Facade inspection robot	Obayashi
169	Facade inspection robot	Takenaka
170	Facade element inspection robot KABEDOHA	Obayashi
171	RIWEA – Robot for the inspection of wind turbine rotor blades	Fraunhofer IFF
172	Bio-inspired facade inspection robot "C-Bot"	Niklas Galler (nr21 DESIGN GmbH)
173	CAFE – Semi-automated cost-effective cleaning system	Universidad Politecnica de Madrid
174	Rail-guided facade-cleaning robot Canadian Crab	Shimizu
175	RobuGlass facade-cleaning robot	Robosoft
176	SIRIUS facade-cleaning robot platform	Fraunhofer IFF
177	Facade-cleaning robots for large glass facades of public buildings (train stations, trade fairs, airports, etc.)	Fraunhofer IFF
178	Rail-guided modular facade-cleaning robot	Department of Mechanical Engineering, Korea University
179	Rail-Guided facade-cleaning robot	Nihon Bisoh
180	Bespoke access solution	SCX Special Projects
181	Modular maintenance and rescue robot system	Komatsu
182	Firefighting robot	Tachi Laboratory, The University of Tokyo
183	Universal firefighting and handling robot	Robotsystem
184	Subsurface ground inspection robot	Tokyu
185	Desk and chair arrangement robot	Fujita
186	Multipurpose travelling vehicle (MTV)	Shimizu
187	Aeration system inspection robot	Fujita
188	WINSPEC air volume measuring robot	Shinryo

the wall. The light weight of the robot, only about 25 kilograms, provides advantages regarding transport and set-up and disadvantages regarding robust operation during unsuitable weather conditions. The robot is able to inspect a facade surface of about 450 m^2/day.

For further information, see Inoue et al. (2009).

System 166: Facade Inspection Robot

Developer: Kajima Corporation

The inspection of tile facades requires usually human workers to be lifted down by gondolas or to work on scaffolding and if done manually is an error-prone work. For this reason, Kajima has developed the "impulse response method", an accurate method for automatic facade inspection. In this method, a perpendicular hammering with a constant pressure is performed on tile surfaces. Tile separation is detected by the maximum amplitude and frequency characteristics of the reflected (and by an on-board microphone recorded) hammering sounds. The system consists of the main

Figure 2.328. Tile facade inspection robot system in action. (Image: Obayashi Corporation.)

robot unit, a rail-guided suspension system located at the rooftop for suspending the main robot unit and moving it horizontally along the facade, the hammering and sound detection system, and a control station located on the ground (Figures 2.329 and 2.330).

Figure 2.329. Testing of facade inspection robot prototype.

Figure 2.330. Real-world application of facade inspection robot. (Image: Kajima.)

Figure 2.331. Façade inspection robots – 1. (Image: Taisei Corporation.)

System 167: Facade Inspection Robot

Developer: Taisei Corporation

Taisei's facade inspection robot (Figure 2.331) was developed to ensure quick and accurate inspection of tiles' condition and their safe adherence to the building. The robot is fixed to the facade using a cable system, suspended from the roof of the building. Installation of the robot does not require any extra preparatory work, as is the case with conventional working platforms or scaffoldings. To reduce the robot's weight and size, the system is divided into several separate parts: (1) the actual robot unit containing the inspection end-effector, (2) a cable suspension system placed at the top and the bottom of the building, and (3) a host computer unit. The actual robot unit contains an activator allowing one to move the unit up or down. Power supply is conducted through the bottom part of the cable suspension system. The actual robot unit is 1150 millimetres wide, 400 millimetres high, and 400 millimetres deep. It weighs approximately 70 kilograms including the lifting motor. The robot has a maximum speed of 6 m^2/minute and can finish an average of 700 m^2/day including preparation time. The inspection end-effector consists of small diagnosis balls tapping on the tiles and a sound analysis unit.

For further information, see Ebihara (1988).

System 168: Facade Maintenance Robot

Developer: Obayashi Corporation

This robot was developed by Obayashi to process the exterior wall tiles of high-rise buildings with balconies (Figures 2.332 and 2.333). The robot consists of a suspension system, a main body frame with end-effectors that allow the robot to adhere safely to the wall/balcony, and a processing unit. General positioning of the system is achieved by the suspension system. Once temporarily adhered to the wall, a part of the main body frame can then be lifted and lowered to move the processing unit inspection unit over the surface area to be inspected. The robot may be equipped with different processing units such as for cleaning and wall tile inspection.

Figure 2.332. Facade maintenance robot system in action. (Image: Obayashi Corporation.)

Figure 2.333. Test of facade maintenance robot prototype. (Image: Obayashi Corporation.)

System 169: Facade Inspection Robot

Developer: Takenaka Corporation

The conventional method of inspecting tile walls can be time consuming, as it requires setting up scaffolding and manual labourers climbing up the facade to inspect each tile. The wall inspection system was developed by Takenaka to automate this task (Figures 2.334 and 2.335). The robot is able to move vertically along

Figure 2.334. Detailed view of facade inspection robot. (Image: Takenaka Corporation.)

Figure 2.335. Facade inspection robot system in action. (Image: Takenaka Corporation.)

Figure 2.336. Facade element inspection robot. (Image: Obayashi Corporation.)

the facade and integrate an end-effector that allows it to inspect tiles by applying and analysing continuous vibrations to the outer surface of the tiles.

System 170: Facade Element Inspection Robot KABEDOHA

Developer: Obayashi Corporation

The facade element inspection robot (Figure 2.336), developed by Obayashi, is used for the automatic inspection of exterior facades. The robot is able to move vertically along the facade and contains an inspection module that uses a camera-based system to analyse the facade and detect irregularities in the facade surface.

System 171: RIWEA – Robot for the Inspection of Wind Turbine Rotor Blades

Developer: Fraunhofer IFF

The RIWEA concept was developed as an alternative to the resource-intensive and downtime-causing inspection of the rotor blades of large wind turbine towers. The robot consists of a square-like frame that houses all components relevant for the operation of the system such as the inspection end-effectors (equipped, e.g., with sensors for inspection by infrared thermography, ultrasonic, cameras, etc.) and the sliding mechanism (Figure 2.340). The two inspection end-effectors move during the inspection process in a constant distance over the blade's surface. This is enabled by a 3-DOF mechanism guided by two rails attached to the bottom of the outer two parts

Figure 2.337. The robot's square-like frame can be lifted up the turbine tower by ropes. (Image: Fraunhofer IFF.)

Figure 2.338. The robot's frame integrates activators and a frame opening/closing-mechanism that allow the robot to climb over from the turbine mast to the rotor blade. (Image: Fraunhofer IFF.)

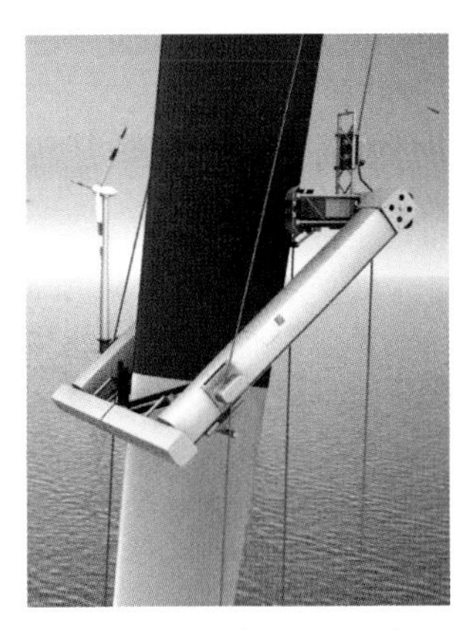

Figure 2.339. Once embracing the rotor blade the robot's frame can be lifted down along it and flexibly adjust to its dimensions. (Image: Fraunhofer IFF.)

of the robot's frame that manipulate the end-effectors. The square-like frame can be lifted up the turbine tower and then down the rotor blade in an automated manner by ropes that are attached to the upper end of the rotor blade and that might be foreseen as elements inbuilt into the wind turbine (Figures 2.337–2.339). The frame

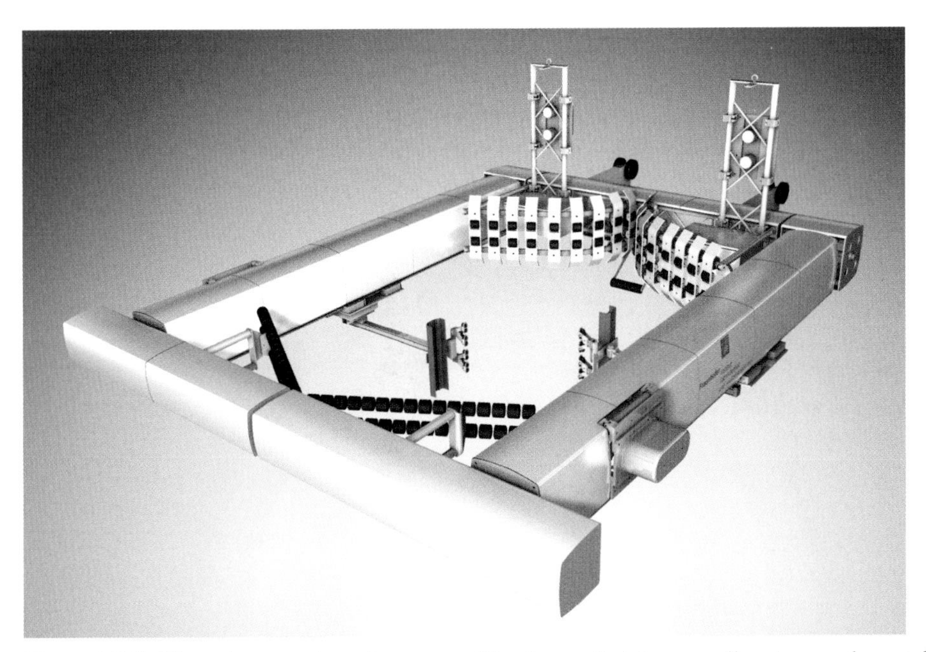

Figure 2.340. The robot consists of a square-like frame that houses all systems relevant for the operation of the system such as the inspection end-effectors (equipped, e.g., with sensors for inspection by infrared thermography, ultrasonic, cameras, etc.) and the sliding mechanism. (Image: Fraunhofer IFF.)

integrates activators and a frame opening/closing mechanism that allows the robot to climb over from the turbine mast to the rotor blade at a certain height. The frame also contains activators that allow the size of the frame to be adjusted to the general dimensions of a rotor blade as well as to the changes in dimensions that occur when being lifted down along the rotor blade.

For further information, see Elkmann et al. (2010).

System 172: Bio-Inspired Facade Inspection Robot C-Bot

Developer: Niklas Galler (nr21 DESIGN GmbH) in cooperation with Prof. Dr. Stanislav Gorb (at that time working for Max Planck Institut for Metal Research, Evolutionary Biomaterials Group)

The C-Bot is a bio-inspired robot concept for the inspection of facades of buildings (Figure 2.341). The main body of the robot can be equipped with sensors such as ultrasonic sensors that allow detection of inconsistencies, damages, and rust inside

Figure 2.341. C-Bot inspecting a concrete facade. (Image: Niklas Galler, nr21 DESIGN GmbH.)

Figure 2.342. Detail of the activation system of the climbing end-effector. (Image: Niklas Galler, nr21 DESIGN GmbH.)

reinforced concrete or cracks in the surfaces facades. The structure of links and joints around the main body allows the robot to move flexibly and omnidirectionally along the facade and around obstacles or corners in the facade. In cooperation with the Max-Planck-Institut and a specialist in functional morphology and biomechanics (Prof. Dr. Stanislav Gorb), the climbing end-effectors were developed (Figures 2.342–2.343). The pads that are part of the end-effectors are inspired by the foot soles of geckos and allow safe adhesion to the facade without external power input on the basis of van der Waals forces. Materials using the gecko approach were developed previously by Prof. Dr. Stanislav Gorb. For this concept study, to create the required adhesion effect on the pad surface of the climbing end-effectors, it was assumed that it would be possible already to scale down the structure of such materials to a nano level.

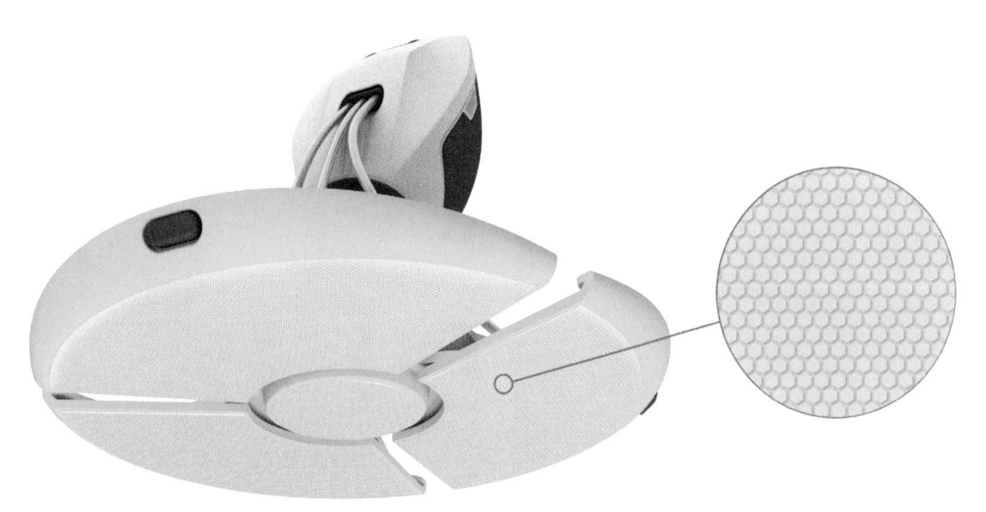

Figure 2.343. The pads that are part of the climbing end-effectors are inspired by the foot soles of geckos and allow safe adhesion to the facade. (Image: Niklas Galler, nr21 DESIGN GmbH.)

Figure 2.344. CAFE – Semi-automated cost-effective cleaning system in action. (Image: Universidad Politecnica de Madrid/Prof. Gambao/Gambao et al. 2008.)

System 173: CAFE – Semi-Automated Cost-Effective Cleaning System

Developer: Departamento de Automatica, Ingenieria Electronica e Informatica Industrial, Universidad Politecnica de Madrid, Prof. E. Gambao

CAFÉ (Figures 2.344–2.346) is a low cost, semi-automatic, and customizable robot kit system that addresses the need for facade cleaning of high-rise buildings

Figure 2.345. Overview of CAFE – Semi-automated cost-effective cleaning system.

Figure 2.346. End-effector of CAFE – Semi-automated cost-effective cleaning system. (Image: Universidad Politecnica de Madrid/ Prof. Gambao/Gambao et al. 2008.)

at reasonable cost. The system consists of a carrier module (installed at the top of the roof and able to lift the main robot module); the main robot module consisting of a kinematic system, a platform module, and a cleaning module; and a control station located on the ground. The kinematic module can be customized to a variety of buildings and can consist of elements such as wheels and/or suction cups that allow the system to be lifted down a facade and during cleaning adhere temporarily to the building. Similarly the cleaning system can be customized. The cleaning system is at a horizontal gantry-type mechanism at the bottom of the main robot module. The platform module is located at the back of the kinematic module and contains system parts that are in general common for each version of the robot. The system can clean facades at a rate of about 200 m^2/hour.

For further information, see Gambao and Hernando (2006) and Gambao et al. (2008).

System 174: Rail-Guided Facade-Cleaning Robot Canadian Crab

Developer: Shimizu Corporation

The facades of high-rise buildings require regular maintenance and cleaning of large glass facades. Shimizu's Canadian Crab (Figure 2.347) glass facade–cleaning robot fully automates the process of class cleaning of a predefined and preprogramed glass surface. The robot system uses three arms, equipped with vacuum suction pads,

Figure 2.347. Rail-guided facade-cleaning robot Canadian Crab. (Image: Shimizu Corporation.)

to travel along and fix itself to the glass facade while cleaning its surface. The robot is supplied via a cable system with the required cleaning liquid.

System 175: RobuGlass Facade-Cleaning Robot

Developer: Robosoft

The Louvre is a modern, glass-based architectural structure that represents a challenge in the cleaning of its glass facade surfaces. Therefore, RobuGlass (Figures 2.348 and 2.349), a facade-cleaning robot, was designed and developed by Robosoft. The company Robosoft was founded in the early 1980s, and in cooperation with other partners, has in particular contributed to the field of mobile robotics by developing robotic solutions for many different fields such as healthcare robots, service robots, defence and security robots, and logistic supply robots. The early prototype of RobuGlass utilized a caterpillar-like locomotion mechanism to move along the glass surface of the pyramid and conduct the necessary cleaning tasks. The latest version consists of a cleaning unit, a crawling system, and a power supply unit. The robot moves over the Louvre's surface using a vacuum system and conducts the cleaning tasks using an end-effector containing a rotating brush for cleaning and a rubber blade for drying. The power supply unit remains on the ground and supplies the mobile cleaning unit with power and water. The robot can move on surfaces with an inclination of up to 90° and conducts cleaning with a speed of up to 25 cm/s. The system allows a significant saving of water as it uses only 4 litres per hour. RobuGlass can be operated automatically remotely controlled by a single worker. The robot has dimensions of

Figure 2.348. RobuGlass in action. (Image: Robosoft.)

Figure 2.349. First prototype of RobuGlass with caterpillar-like locomotion mechanism. (Image: Robosoft.)

1200 × 900 millimetres and weights only about 55 kilograms. The robot is currently scheduled to clean once every 3 weeks.

For further information, see Robosoft (2016).

System 176: SIRIUS Facade-Cleaning Robot Platform

Developer: Fraunhofer IFF

The SIRIUS platform incorporates a cleaning end-effector that travels along a linear axis that is mounted on the bottom of the platform's main body. The elastically supported end-effector uses rotating brushes in combination with water to clean the windows. The injected water is absorbed by the brushes, filtered, and returned to an on-board water storage tank. The robot can either be started from atop the roof or installed to the trolley cables by operators on the ground level. After refilling of the water storage tank, the robot is placed on the facade in the starting position for the cleaning procedure. Based on the SIRIUS platform a series of cleaning robots

Figure 2.350. SIRIUS facade-cleaning system version 3. (Image: Fraunhofer IFF/ Bernd Liebl.)

Figure 2.351. SIRIUS facade-cleaning system version 5. (Image: Fraunhofer IFF/ Bernd Liebl.)

was developed, each customized to a certain extent to the building on which the robot was supposed to operate (Figures 2.350–2.353). It is planned that in the future additional end-effectors will be developed so that based on the platform also other service robots (e.g., for inspection) can be developed. Robots based on the SIRIUS

Figure 2.352. SIRIUSC facade-cleaning system version 4. (Image: Fraunhofer IFF/Bernd Liebl.)

Figure 2.353. SIRIUS facade-cleaning system version SIRIUS ZV for Fraunhofer Headquarter in Munich. (Image: Fraunhofer IFF/Bernd Liebl.)

platform can clean at a rate of up to 75 m^2/hour, including the time needed to move the robot along the facade. The robots are equipped with infrared sensors to detect obstacles within the facade and to optimize the cleaning procedure.

For further information, see Hortig et al. (2001).

System 177: Facade-Cleaning Robots for Large Glass Facades of Public Buildings (Train Stations, Trade Fairs, Airports, etc.)

Developer: Fraunhofer IFF

Fraunhofer IFF developed a series of gondola-type facade-cleaning robots for large glass facades of public buildings (train stations, trade fairs, air ports, etc.). One

Figure 2.354. Façade cleaning robot for Leipzig's new Exhibition Centre. (Image: Fraunhofer IFF.)

robot (Figure 2.354) was developed for the maintenance of the Glass Hall at the Leipzig Trade Fair (25,000 m^2 glass facade). To keep glass transparent the glass facade must be cleaned frequently. The system consists of a central module travelling along the highest level of the glass roof from which the main robot units (containing the cleaning end-effector) are then suspended: one on the north side and one on the south side of the building. The central module also supplies water to the suspended cleaning robots. The end-effectors utilize a roller brush mechanism to clean the main surfaces of the glass and retractable, dexterous disc brushes for the cleaning of areas around the glass mounts. The size of the system (W 150 cm × H 30 cm) and its weight (250 kg) are fully adapted to the facade's exoskeleton-like steel structure, dimensions, and load-bearing capacity. Another robot system (Filius Toni; Figure 2.355) that can be controlled by an operator in a similar way from a central module travelling along the highest level of the glass roof was developed for the Berlin Central Train Station. The smooth surface of the glass facade required here only a relatively simple roller brush mechanism.

For further information, see Elkmann et al. (2002).

Figure 2.355. The facade cleaning robot Filius Toni was adapted to the roof of the Berlin Central Train Station. (Image: Fraunhofer IFF/Justus Hortig.)

System 178: Built-In Rail-Guided Building Facade Maintenance Robot

Developer: Department of Robot Engineering, Hanyang University, Ansan Korea, Prof. Chang Soo Han, PhD, Min Sung Kang, PhD, Seunghoon Lee, Yong Seok Lee, Sang Ho Kim, and Myeong Su Gil; Department of Mechanical Engineering, Korea University, Seoul, Korea, Prof. Dae Hie Hong, PhD, Sung Min Moon, Sung Won Kim, So Ra Park, and Jae Myung Huh

This rail-guided facade-cleaning robot was developed to address the increasing need for building facade cleaning as the number of high-rise buildings is continuously increasing. In contrast to most facade-cleaning systems that use a carrier (or gondola), this robot uses a rail-guided system for horizontal and vertical movements along the building (Figures 2.356 and 2.357). It uses a wheel moving mechanism for horizontal movement once the cleaning task at a certain level of building is finished. The robot moves vertically by using a winch system. This built-in facade maintenance system also has a docking station, with docking connector, for the robot to receive material supplies from the building and uses an electric valve system to control it. By using this robot, maintenance workers are no longer required to work on difficult-to-access facade areas of building. In manual facade cleaning, a cleaning worker on a suspended platform or gondola must exert some force on window panels for cleaning purposes, but this robot's horizontal and vertical guiderails impose the load on the rail rather than the window.

Figure 2.356. Cleaning work using the horizontal robot. (Image: Department of Robot Engineering, Hanyang University/Department of Mechanical Engineering, Korea University.)

Figure 2.357. Cleaning work using the vertical robot. (Image: Department of Robot Engineering, Hanyang University/Department of Mechanical Engineering, Korea University.)

System 179: Rail-Guided Facade-Cleaning Robots

Developer: Nihon Bisoh Co., Ltd.

Nihon Bisoh Co., Ltd. is a Japan-based company that provides robotic, automated, semi-automated, and nonautomated solutions for the inspection, cleaning, and refurbishing of the facades of large buildings. The company was established in 1966, and today has about 500 employees. The company provides gondola-type systems that are suspended from the ceiling and move along the facade vertically as well as rail-guided systems that process the facade segment by segment in a horizontal manner. Nihon Bisoh provides standard modules, customized versions, and a full range of planning and implementation services. In most cases Nihon Bisoh's systems are structurally and in terms of design fully integrated with the building (e.g., rails on which the cleaning robots are operated are in many cases an integral part of the facade; Figures 2.358–2.361), allowing them to be equipped with a variety of

Figure 2.358. Carrier system moves the cleaning module vertically from one floor level to another. (Image: Nihon Bisoh Co., Ltd.)

Figure 2.359. Cleaning module release from the carrier module starts operation horizontally along the rails integrated into the facade. (Image: Nihon Bisoh Co., Ltd.)

Figure 2.360. System in action on the facade of a high-rise building/Landmark Tower. (Image: Nihon Bisoh Co., Ltd.)

Figure 2.361. System in action on the facade of a high-rise building/Crystal Tower. (Image: Nihon Bisoh Co., Ltd.)

end-effectors that can be retracted or stored invisibly when not used in dedicated areas in the building.

System 180: Bespoke Access Solution

Developer: SCX Special Projects Ltd.

SCX develops and installs customized cleaning and maintenance solutions for large buildings. In the case of The Gherkin (location: London; architect: Foster + Partners), because of its complex structure and large glass surface, SCX developed a partly automated system that is almost fully integrated into the building (Figure 2.362). On level 36 a ring of rails (architecturally and structurally a part of the building) serves as the basis for the operation of three building maintenance units (BMUs). A BMU is a mobile base unit that can be operated and driven along the rail system around the building. Two of these BMUs are equipped with a horizontally

Figure 2.362. BMUs of the bespoke access solution (engineered and installed by SCX Special Projects Ltd.) in action. (Image: SCX Special Projects Ltd.)

oriented telescopic arm from which end-defectors for the cleaning of the floors under the rail ring can be lowered and positioned. One BMU is equipped with a five-stage hydraulic manipulator with a reach of up to 33 metres for the carrying of end-effectors for the cleaning of the area above the rail ring. End-effectors are exchangeable and are available in various versions (containing cleaning equipment, working platforms, glass replacement equipment, etc.). When not in use the BMUs (including the manipulators and end-effectors) can be retracted through a gate in the fence into a dedicated area in the building where they can be stored and prepared for the required maintenance or cleaning task.

System 181: Modular Maintenance and Rescue Robot System

Developer: Komatsu Ltd.

This modular maintenance and rescue robot system (Figure 2.363) was developed by Komatsu to automate tasks such as fence inspection and window cleaning as well as to help rescue people in the event of an emergency. The robot consists of three main parts: (1) the carrying unit located at the top of the roof that moves the whole system horizontally and vertically, (2) a vertical frame that guides the working unit, and (3) the working unit itself. Depending on the building and the tasks to be performed, the system can be equipped with customized working units. In case of an emergency the working unit can be replaced by a rescue firefighting unit.

Figure 2.363. 3D model of maintenance and rescue robot system.

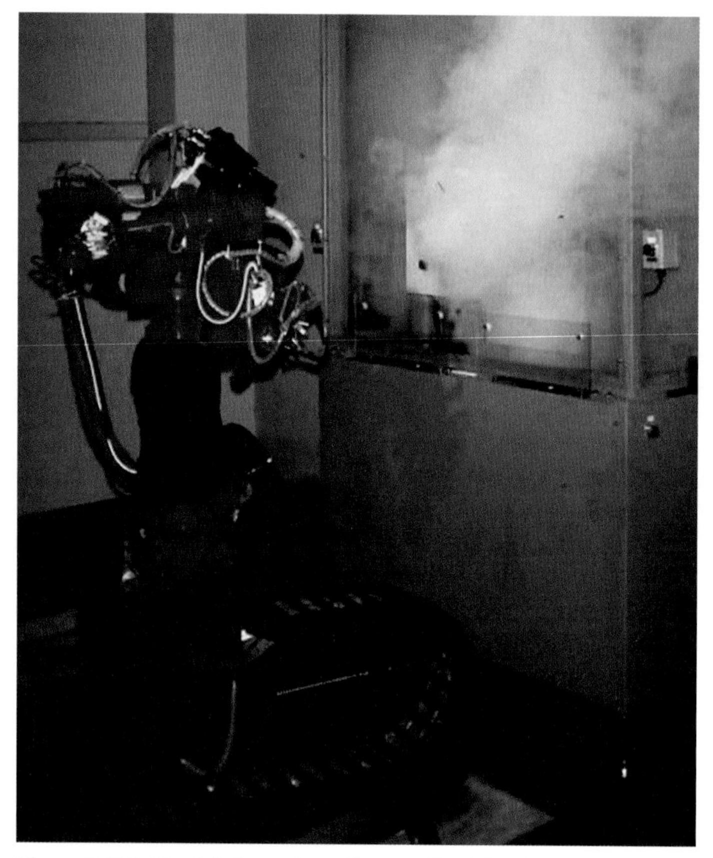

Figure 2.364. Firefighting robot. (Image: Prof. Bock at Tachi Laboratory, the University of Tokyo.)

System 182: Firefighting Robot

Developer: Tachi Laboratory, the University of Tokyo

This firefighting robot (Figure 2.364) was developed by a Japanese research lab. The compact system can either be brought to the site in case of a fire or it can be foreseen as an integral part of a building that becomes active in case of a fire. The extinguishing end-effector is positioned and oriented by a manipulator allocated atop a tracked wheel system that allows it to move flexibly over uneven terrain. The mobile base also contains compartments for storing the extinguisher materials.

System 183: Universal Firefighting and Handling Robot

Developer: ROBOTSYSTEM

ROBOTSYSTEM's universal firefighting and handling robot is a robot on a mobile tracked base equipped with manipulator with 5 DOF (Figure 2.365). The robot can be remotely controlled and its end-effector contains mechanisms for both grasping (removing elements from the ground, and for screwing of grasped objects; Figure 2.367) and fire extinguishing (by water gunning; Figure 2.366). The

Figure 2.365. Overview of universal firefighting and handling robot. (Image: ROBOTSYSTEM.)

fire-extinguishing medium is guided to the end-effector inside of the manipulator. With the end-effector, the robot can lift subjects weighing 400 kilograms. In another research project, the development will result in a mechanism enabling transformation of the end-effector into a penetrating aggregate that allows drilling of holes into concrete walls and simultaneously spraying water behind the wall for firefighting purposes. The total system weighs 4000 kilograms and is equipped with a stereovision camera system; the maximal reach of the manipulator equals 3.8 metres; the manipulator and the end-effector are partly hydraulically activated; and the tracked mobile platform is equipped with a Kubota V3800-DI-T-E3 diesel engine. The dimensions of the system are 3.1 metres (L) × 2.8 metres (W) × 2.9 metres (H).

Figure 2.366. End-effector in water gunning mode. (Image: ROBOTSYSTEM.)

Figure 2.367. End-effector in element grasping mode. (Image: ROBOTSYSTEM.)

Figure 2.368. Subsurface ground inspection robot. (Image: Tokyu Construction Co., Ltd.)

System 184: Robot Ladybird for Nondestructive Ground Subsurface Surveying

Developer: Tokyu Construction Co., Ltd.

The Ladybird robot (Figure 2.368) was developed by Tokyu Construction for the nondestructive surveying of the subsurface area of a ground area (e.g., roads) as a premeasure of maintenance planning. The robot consists of a mobile platform equipped with an end-effector with an electromagnetic subsurface radar. This end-effector contains antennas for the emission of electromagnetic energy and the sensing and interpretation of the reflections allow detection of the presence, size, and location of cracks, hollow areas, and objects. The robot makes it possible to survey large areas and produce detailed images of the conditions under the surface. The robot is available in two variants, one for the surveying of the subsurface area up to a depth of 1 metre and one for depth up to 3 metres.

System 185: Desk and Chair Arrangement Robot

Developer: Fujita Corporation

This proposed robot system is designed to arrange various types of desks and chairs in an individual room (Figure 2.369) according to the room's intended purpose, and after their use, rearrange, remove, and/or store them. Each robot consists of a

Figure 2.369. Desk and chair arrangement robot system in action. (Image: Fujita Corporation.)

mobile platform and a forklift–like mechanism. The robot is able to insert a forklift-like mechanism under system desks and chairs and lift and transport them to the predetermined locations. Part of the system is also an automated warehouse with racks for storing desks and chairs with which the robot can interact and in which it can place the desks and chairs. The system correctly identifies and arranges chairs based on a vision system installed as global sensor system to the ceiling and RFID chips integrated in each chair.

For further information, see Kumita et al. (1992).

System 186: Multipurpose Travelling Vehicle (MTV)

Developer: Shimizu Corporation

The multipurpose travelling vehicle (MTV; Figure 2.370) was developed by Shimizu originally to automate the concrete-slab finishing work. However, as the system could, because of a modular approach, be equipped with other end-effectors also, an exchangeable floor cleaning end-effector was developed as well. The system (dimensions: L 1.35 m × W 0.70 m × H 0.90 m) consists of a mobile platform (180 kg) to which either a cleaning module (about 55 kg) or a grinding module (90 kg) can be attached. The system requires a worker to set up the system before it can work autonomously in a predefined area. The robot moves first throughout the work area and maps the environment and then conducts the work process autonomously.

For further information, see Kangari and Yoshida (1989).

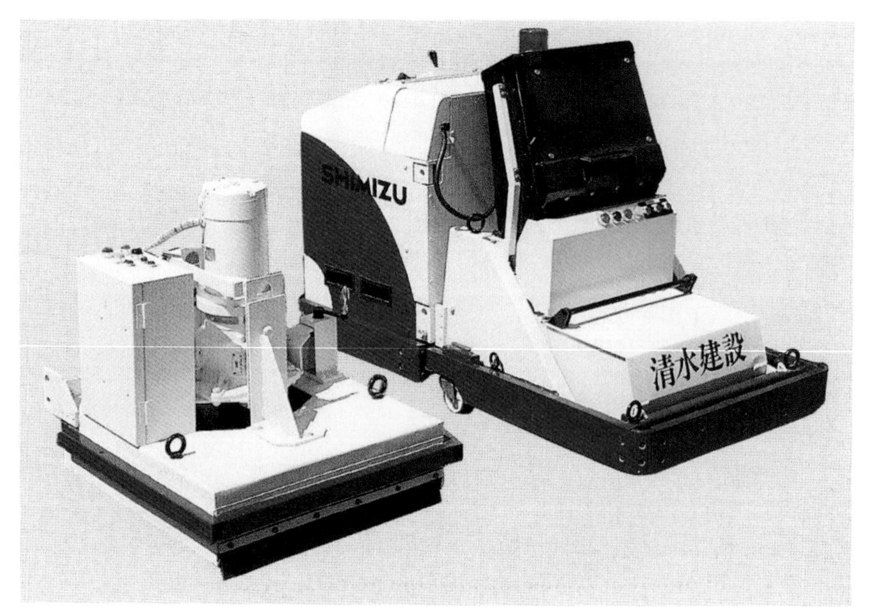

Figure 2.370. MTV-1 mobile base equipped with floor grinding module. (Image: Shimizu Corporation.)

System 187: Aeration System Inspection Robot

Developer: Fujita Corporation

The aeration system inspection robot (Figure 2.371), developed by Fujita, is used for the automatic inspection of aeration ducts. To ensure the efficiency of any HVAC

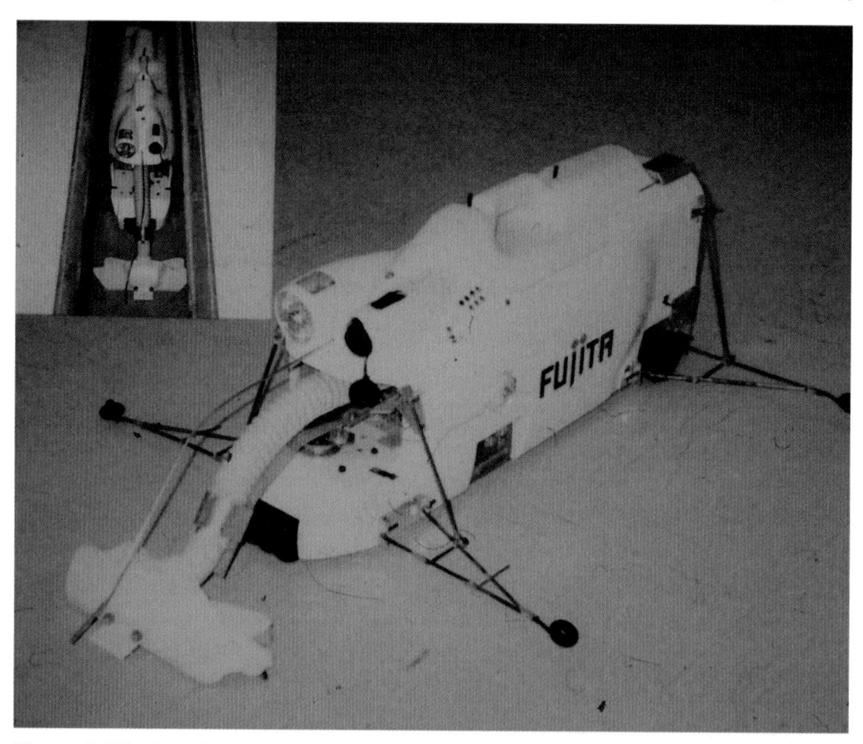

Figure 2.371. Aeration system inspection robot. (Image: Fujita Corporation.)

systems (heating, ventilation, air conditioning), it is important to confirm that the aeration infrastructure (air ducts) is airtight. It is also important to identify and locate any defects that may be present in the ducts. The system is able to navigate autonomously through aeration ducts and conduct inspection.

System 188: WINSPEC Air Volume Measuring Robot

Developer: Shinryo Corporation

WINSPEC (Figures 2.372–2.374) is a device designed to measure the volume of air that comes out of the aeration system openings. There are many models of the WINSPEC system that were developed for different types of buildings; the first prototype was completed in 2000. Based on the measurements taken by this device, the required actions can be taken to adjust the aeration control system or plan for repairing and renovation. The robot is easy to use with a user-friendly interface in the form of a touch screen panel. A microprocessor takes the command to activate the air velocity sensors in the end-effector. The end-effector then takes multiple measurements of the air velocity at different points through the aeration system opening.

The figure shows the following different models and end-effectors of the system:

1. WINSPEC-2: Standard model designed for buildings with ceiling heights of up to 3 metres
2. WINSPEC-i: Variation of the standard model; designed for buildings with ceiling heights of up to 4.5 metres

Figure 2.372. Prototype version of WINSPEC. (Image: Shinryo Corporation.)

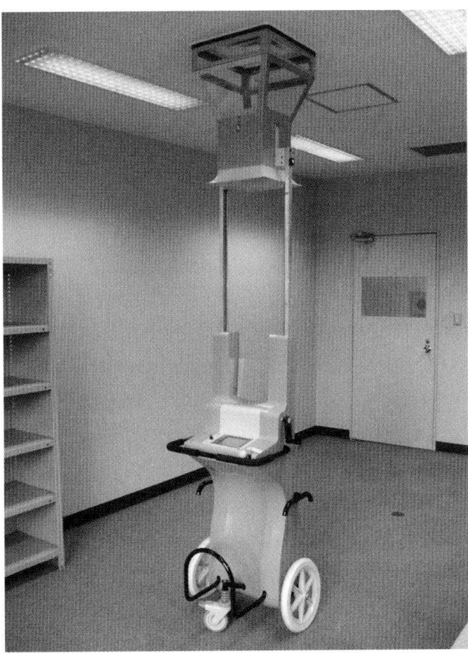

Figure 2.373. Commercialized WINSPEC version. (Image: Shinryo Corporation.)

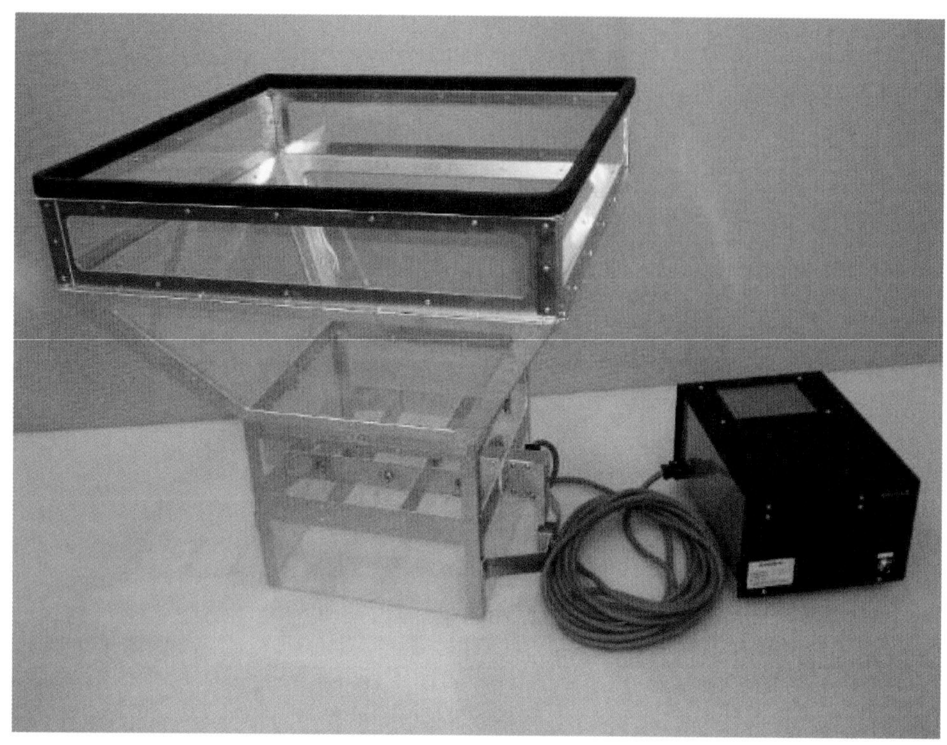

Figure 2.374. WINSPEC end–effector. (Image: Shinryo Corporation.)

3. WINSPEC-P: End-effector of the WINSPEC robot system able to be used as a standalone system; used in situations in which the location to be inspected cannot be accessed by the other models

For further information, see Tanaka et al. (2006).

2.24 Renovation and Recycling Robots

Renovation and recycling projects are usually labour intensive and difficult. In 1997, the German construction industry experienced an interesting turning point: the construction volume associated with renovation and modernization exceeded the construction volume of new constructions. Today, the discrepancy is even larger than before and it is considered certain that this trend will continue. In Japan, where buildings usually have shorter life cycles, the trend is also heading towards higher volumes of renovation to save resources. On a worldwide scale it can observed that the demand for more and more energy-efficient buildings becomes a driver for renovation and building disassembly. The automation of renovation, disassembly, and recycling operations, on the one hand, promises tremendous improvements in labour productivity, but on the other hand, it demands a relatively high flexibility of the robot systems compared to other STCR categories. Robots in this category include systems for the demolition and disassembly of the building's structural element and interior; robots for the on-site crushing and recycling of material; preparation and concrete surface removing robots for surfaces such as floors, walls, and facades; and finally yet importantly robots used to assist in and automate the dangerous asbestos removal process.

Renovation and Recycling Robots		
189	DXR 310 demolition robot	Husqvarna
190	On-site material recycling robot Garapagos	Komatsu
191	Concrete surface removing robot I	Shimizu
192	Concrete surface removing robot II	Shimizu
193	Automatic chimney cleaning and brick dismantling equipment	Toda
194	Facade delamination robot Jet Scrapper	Takenaka
195	Post-building-demolition/post-disaster cleanup robot	Tmsuk
196	Water blasting robot	Kajima
197	Carbon fibre layer application robot	Obayashi
198	Asbestos removal, collecting, and packaging robot	Taisei
199	ERO concrete recycling robot	Ömer Hacıömeroglu
200	Robot for the dissolving and removal of asbestos coating material sprayed on steel components	Takenaka Corporation funded by New Energy and Industrial Technology Development Organization (NEDO)

System 189: DXR 310 Demolition Robot

Developer: Husqvarna AB

The DXR 310 is a remote-controlled robot that can be used for a variety of tasks in the context of demolition (Figure 2.375). The robot is compact and with its manipulator in folded position can be transported or driven through doors and other

Figure 2.375. Several robots used in parallel to demolish parts of a building. (Image: Husqvarna AB.)

openings (width: 0.78 m). The manipulator (range: 5.5 m; telescopic beam) can be equipped with a variety of end-effectors such as demolition hammers, steel cutters, compact crushers, and buckets. The robot is tele-operated remotely over a control panel with joysticks that can be worn by the operating construction worker. The robot can optionally be equipped with protective elements and additional cooling elements for operation in environments where high temperatures are expected and it provides a power of 22 kilowatt. It is equipped with tracks for movement and a system that allows stable fixation of the robot on the ground in the desired position and shovelling away of the demolished material.

System 190: On-Site Material Recycling Robot Garapagos

Developer: Komatsu Ltd.

Komatsu developed this robot (Figure 2.376) for the on-site and closed loop reprocessing of construction waste. For example, large concrete pieces that are generated during the demolition of a building can be crushed and pulverized on site and thus reused directly as basic material for on-site production of (cement-based) building elements. The robot system consists of a tracked mobile platform, a processing unit mounted on top of the platform, and a cone-shaped end-effector into which the waste material can be placed (e.g., by an excavator). The system can be used as a stand-alone system on conventional construction sites where excavators can transport the material to the cone-like intake end-effector. It can also be used as a networked system within an automated construction site where material is continuously provided by an overhead manipulator or other automated logistics systems.

Figure 2.376. On-site material recycling robot, Komatsu Ltd.

Figure 2.377. Concrete surface removing robot I system in action. (Image: Shimizu.)

Figure 2.378. Concrete surface removing robot I end-effector. (Image: Shimizu.)

System 191: Concrete Surface Removing Robot I

Developer: Shimizu Corporation

This robot (dimensions: 1.80 m diameter × 1.30 m H; weight: 650 kg) was developed for the processing (e.g. cleaning, removal of concrete surfaces) of large concrete ground surfaces (Figures 2.377 and 2.378). During dam construction and also during the construction of large buildings, thick layers of concrete have to be poured and in many cases the new layer must form good bonds with an existing old or previously poured layer of concrete. The robot can move omnidirectionally on such types of concert surfaces and is used to prepare the existing concrete surface. The robot consists of a robotic mobile platform with four drive wheels allocated in the middle able to provide omnidirectional on-site mobility, set of three hydraulically activated robotic manipulators that can be equipped with a variety of brush devices and discs, an on-board power generator, and a remote control panel that can be worn by the operator tele-operating the system. The robotic manipulators are able to automatically regulate/adjust the pressure depending on the task, end-defector, and surface condition. The robot can process 1.5-metre work area strips, moves with a speed of 14 m/minute, and is able to process up to 300 m^2/hour.

For further information, see Council for Construction Robot Research (1999, pp. 68–69).

System 192: Concrete Surface Removing Robot II

Developer: Shimizu Corporation

This robot was developed for the cleaning, removal, and treatment of concrete surfaces of large areas (Figure 2.379). During dam construction, thick layers of concrete are poured and the new layer must form good bonds with the existing one. The robot can move in one direction and is used to prepare the existing concrete surface and make it ready to develop a strong bond with the new concrete to be poured. The robot consists of a mobile platform and a large end-effector unit mounted to its front end containing a device to cut and clean the concrete surface.

Figure 2.379. Concrete surface removing robot II system in action. (Image: Shimizu.)

System 193: Automatic Chimney Cleaning and Dismantling Robot

Developer: Toda Corporation

The system was developed to automate the process of cleaning and dismantling of the inner brick-based part of chimneys (Figure 2.380). The system is lifted down a chimney and first cleans and then dismantles the inner brick layers. The system (height: 3.8 m; weight: 1.0 tons) consists of a storage jig in which the robot is transported and that is positioned at the top of the chimney and is able to lift down the robot, the robot's main body, two sets of four length-adjustable arms extending from the main body that guide the robot though the chimney, two end-effectors (one for cleaning and one for brick dismantling), water supply and recycling system placed on the ground, and a control work-station placed on the ground. The system is positioned by a crane on the top end of the chimney and from there the main body is suspended from the transporting jig automatically. The spraying nozzle end-effector is able to spray cleaning liquid with a strong pressure to the inside of the chimney. The working area of the spring end-effector can be automatically adjusted according to the diameter of the chimney. Similarly the descending speed of the system is adjusted according to the level of contamination and the chimney's structure. The brick removing end-effector is mounted to the lower sets of arms and dissolves the bricks segment by segment from the chimney. The equipment is tele-operated from the control workstation. The system is able to process chimneys with diameters ranging from 0.88 to 3.2 metres and can process up to 200 m^2/day.

Figure 2.380. Automatic chimney cleaning and dismantling robot system in action. (Image: Toda Corporation.)

For further information, see Nishiyama and Kobayashi (2006).

System 194: Facade Delamination Robot Jet Scrapper

Developer: Takenaka Corporation

The Jet Scrapper (dimensions: W 4.28 m × H 3.01 m × D 0.9 m; weight: 1300 kg) was developed to automate parts of the process of refurbishing the exterior walls of large buildings (Figure 2.381). The refurbishing of exterior walls requires removal of the previously existing coating material before applying a new coating or additional insulation or other components. Traditional methods of dissolving ("delaminating") the existing material from a wall (e.g., grinding, water- or sand-blasting by a worker) lead to the uncontrolled dispersal of waste material and an adverse effect on workers and the surrounding environment. This Jet Scrapper utilizes a high-pressure water jet end-effector (Figure 2.382) to dissolve the existing material from the wall. A high suction vacuum removes the waste material and water from the jet. The system is lifted down along the wall to be processed and is temporarily affixed to the wall through vacuum suction end-effectors. The system is operated remotely from a control workstation on the ground by a single operator and is able to process up to 50 m^2/ hour.

For further information, see Council for Construction Robot Research (1999, pp. 254–255).

Figure 2.381. Facade delamination robot Jet Scrapper system in action. (Image: Takenaka Corporation.)

Figure 2.382. Facade delamination robot Jet Scrapper end-effector. (Image: Takenaka Corporation.)

Figure 2.383. Post-building demolition/post-disaster cleanup robot system in action.

Figure 2.384. Front view of post-building demolition/post–disaster cleanup robot system.

System 195: Post-Building Demolition/Post–Disaster Cleanup Robot

Developer: Tmsuk Corporation

Tmsuk is a small robot builder from Kyushu that continuously develops novel ideas and concepts for robot applications and systems and, in most cases, realizes them as pre-series and prototypes to demonstrate their feasibility. Tmsuk's cleanup robot (Figure 2.383–2.385) was developed to remove components, material, and

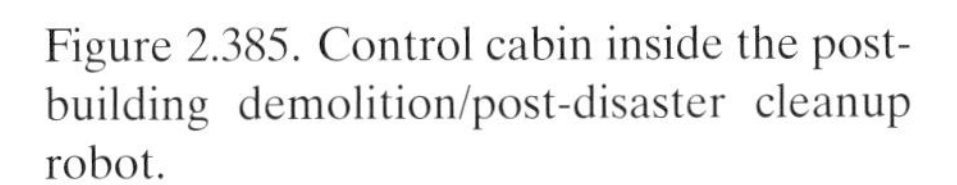

Figure 2.385. Control cabin inside the post-building demolition/post-disaster cleanup robot.

Figure 2.386. Water blasting robot. (Image: Kajima Corporation.)

waste after the demolition/disassembly of buildings or after disasters. It has two manipulators, each equipped with large end-effectors that can grasp medium to large sized elements. The manipulators are attached to a mobile tracked platform that has a steel blade at the front to remove smaller materials from the navigation path and integrates also a control cabin for the operator. A steel cage with a metal frame is installed in the front of the robot to ensure the safety of the operator. Further versions of the system will allow for tele-operation as well as for partly automated material removal.

System 196: Water-Blasting Robot

Developer: Kajima Corporation

Kajima's green water blasting robot (Figure 2.386) is used to process different kinds of surfaces. The robot consists of a mobile platform equipped with a control cabin and a manipulator equipped with a set of water blasting nozzles. The position and orientation of the nozzles are controllable so that the effect of the water blasting on the surface can be customized and adjusted.

System 197: Carbon Fibre Layer Application Robot

Developer: Obayashi Corporation

Tower-like structures are vulnerable structures with respect to earthquakes and other influences (weather, exhaust fumes, etc.) of the surroundings. The carbon fibre layer application robot (Figure 2.387) was developed by Obayashi to apply an additional layer with a carbon-fibre–based material to existing reinforced concrete structures to refurbish them and improve their resistance. The system consists of two winch-type lifting mechanism (placed on top of the structure) and a suspended ring-shaped mechanism that moves vertically along the outer surface of the structure

Figure 2.387. Automatic carbon fiber wall wrapper. (Image: Obayashi Corporation.)

(lifted by the winch-type lifting mechanism). The ring shaped mechanism consists of two rings inside each other; the outer ring revolves along the inner ring, which contains the end-effectors that wrap the carbon-fibre–based material around the structure.

System 198: Asbestos Removal, Collecting, and Packaging Robot

Developer: Taisei Corporation

The removal of asbestos used extensively as a building material in the 1970s is a challenge when buildings are refurbished or deconstructed. In particular, when processed on the construction site asbestos releases small fibres in the micrometre scale into the air, which when inhaled by workers can cause serious health problems. Taisei's asbestos removal robot, therefore, tries to robotize with the described system the entire process from removal of the material from the ceiling, wall, or column (detachment), to the collection of the material from the floor, and then to the pressing and transport ready packaging of the hazardous material. To cover the necessary process chain the robot system consists of three subsystems. The control equipment (subsystem 1) consists of a monitoring robot and a remote control station and ensures that construction workers do not get in direct contact with the hazardous material. The asbestos removal robot consists of a mobile platform, a manipulator (Figure 2.388), and an asbestos detachment end-effector (detachment tool; Figure 2.389). The end-effector uses a high-pressure water jet system to remove the asbestos from

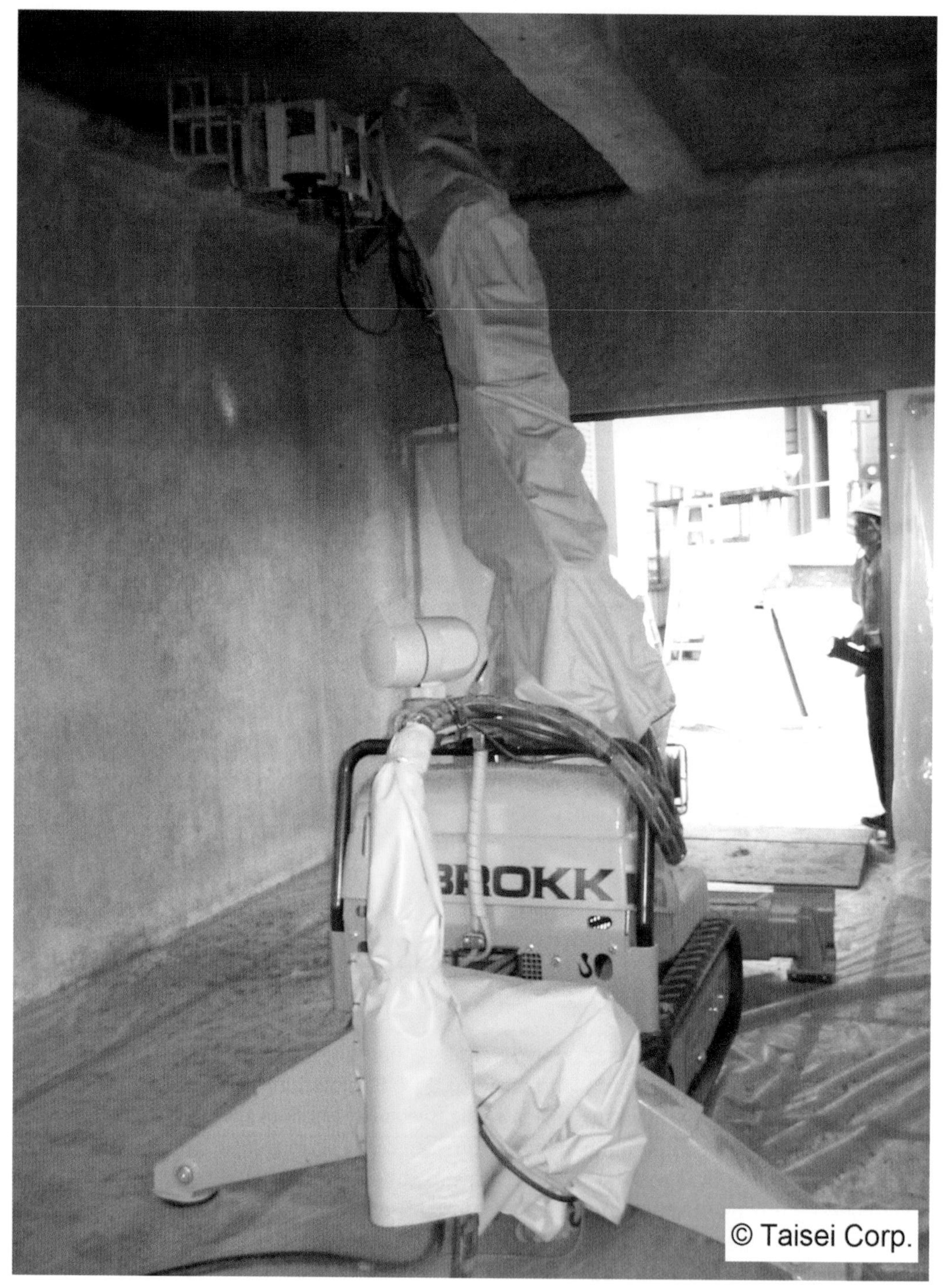

Figure 2.388. Asbestos removal robot. (Image: Taisei Corporation.)

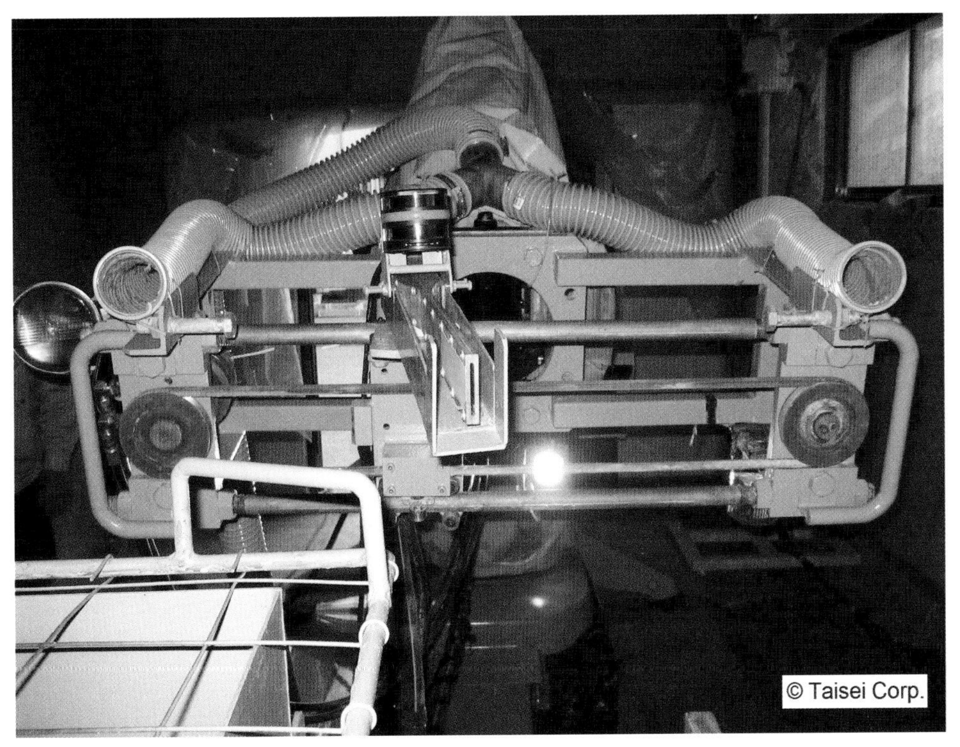

Figure 2.389. Asbestos detachment end-effector. (Image: Taisei Corporation.)

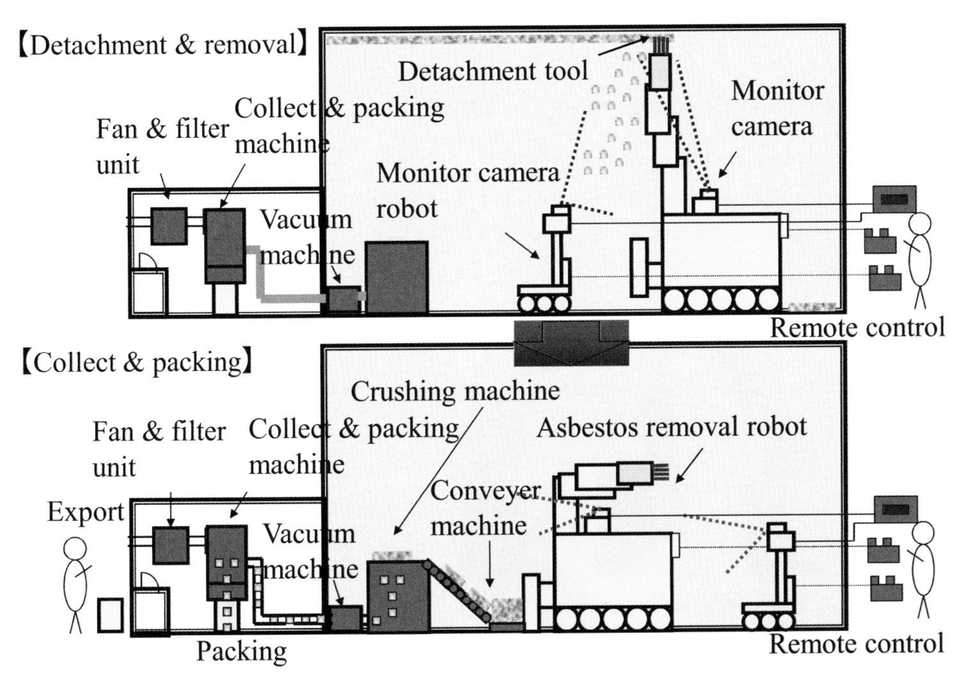

Figure 2.390. Schematic representation of asbestos removal system. (Image: Taisei Corporation.)

Figure 2.391. Visualization of Ero deconstructing a building on site. (Image: Ömer Hacıömeroglu.)

the ceiling, wall, or column. The detached material falls to the floor and is moved by the asbestos removal robot to the collecting and packing system (subsystem 3). This subsystem transports the asbestos over a conveyor and a vacuumed machine into a pressing and packaging system (Figure 2.390).

System 199: ERO Concrete Recycling Robot

Developer: Ömer Hacıömeroglu

ERO (Figures 2.391–2.393) is a robot that can be used to disassemble/remove existing concrete structures and it was designed by the Turkish product and industrial

Figure 2.392. Three phases of Ero (working, mobile, and parking). (Image: Ömer Hacıömeroglu.)

Figure 2.393. Ero and the aggregate packages. (Image: Ömer Haciömeroglu.)

designer Ömer Haciömeroglu. The robot consists of a mobile platform with a wheel system that allows for omnidirectional movement, a manipulator, an end-effector, a removal and packaging unit, and a remote control unit. The main unit of the robot allows the robot to be folded into a compact unit (L 130 cm × W 160 cm × H 180 cm) that can easily be transported to and on the construction site. The end-effector integrates a high water jet system for concrete detachment and a suction system that directly removes the cut off material through the robot main body into the removal and packaging unit where a so-called centrifugal decanter technology separates the concrete aggregate from the water. The filtered water is then fed back to be reused in the deconstruction process. The separated concrete aggregate could then be directly packed into industrial, standard transport bags.

System 200: Robot for the Dissolving and Removal of Asbestos Coating Material Sprayed on Steel Components

Developer: Takenaka Corporation, funded by New Energy and Industrial Technology Development Organization (NEDO)

The system is able to automate the removal of asbestos material from steel beams. The robot is tele-supervised/operated and recognizes partly automatically the location/shape of the steel beam to be processed and follows the steel beam over its length. The system (Figures 2.394–2.396) consists of a mobile robotic platform, a lifting mechanism with a lifting table, a 7-DOF robotic manipulator (Motoman), an asbestos removal end-effector, and an external stationary system that is able to suck the asbestos away through a hose and transfer (blow) it into closed compartments (bags). The removal end-effector consists of a rotary steel wire–based brush system and a vacuum suction system that sucks away the removed material directly after it was dissolved from the steel component so that no asbestos can escape to the surroundings. The system reduces both the labour resources required and the impact of asbestos on the surrounding and the workers. From the lifting mechanism a

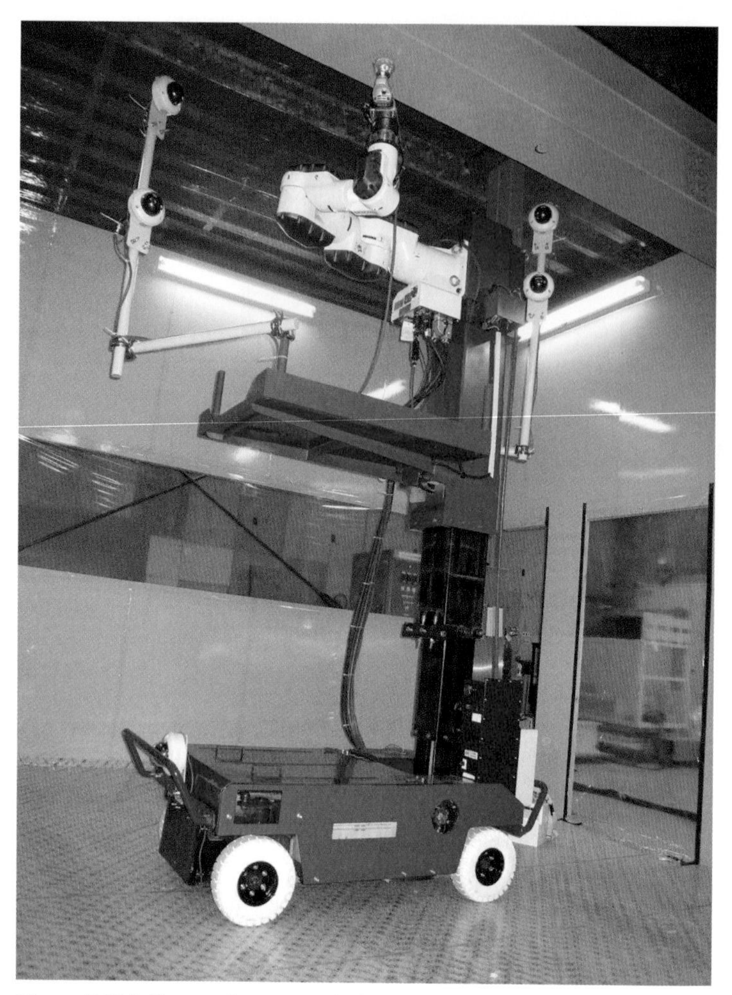

Figure 2.394. System for removal of asbestos on steel components in action. (Image: Takenaka Corporation.)

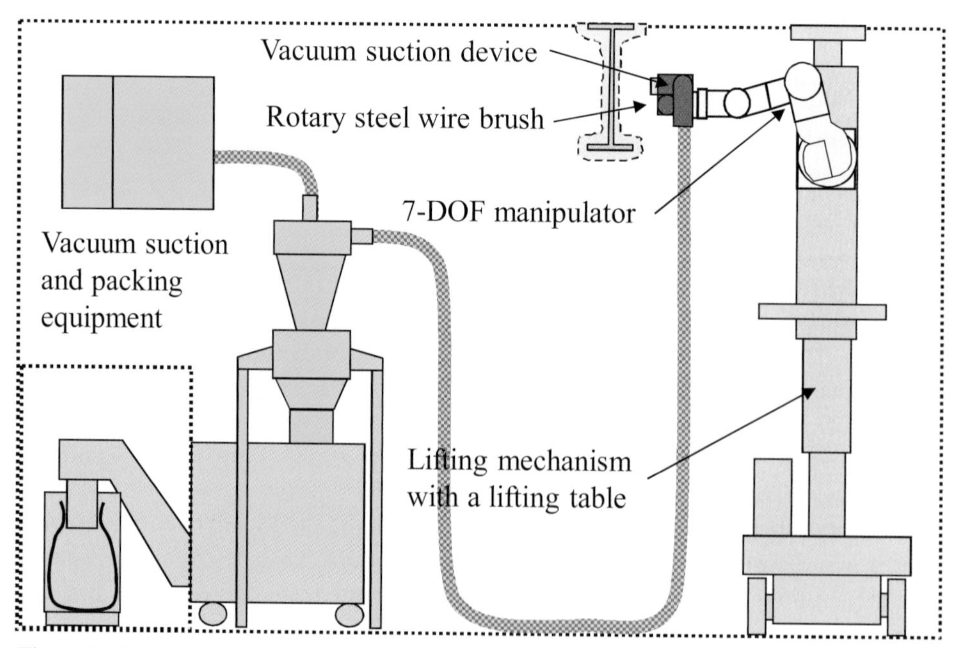

Figure 2.395. Schematic representation of asbestos removal system. (Image: Takenaka Corporation.)

Figure 2.396. Asbestos removal robot. (Image: Takenaka Corporation.)

telescopic arm can extend to the ceiling once the robot is positioned to hold the robot accurately in place during asbestos removal. The system can work in rooms with a height ranging from 1.8 metres to 5.5 metres.

The information and images presented in this outline are based on results obtained from a project commissioned by the New Energy and Industrial Technology Development Organization (NEDO). For further information, see Arai et al. (2010).

3 Transition and Technological Reorientation towards Integrated On-Site Manufacturing

After the evolution of industrialized and automated large-scale prefabrication (LSP) during the 1970s (see **Volume 2**), and the development of robots substituting tasks on the construction site from 1978 onwards (single-task construction robots [STCRs]; see Chapter 2), from 1985 onwards, concepts emerged that envisioned automated site factories. Those concepts combined building component manufacturing technology (processing of prefabricated low-level to medium-level components instead of parts to reduce complexity on-site), STCRs (mainly for interior finishing operations), and moving or stationary site factories that were able to assemble the building's main structure (such as steel frames or concrete structures) almost automatically (see **Volume 4**). As those site factories not only automated parts of the construction process but also "integrated" prefabricated component technology, STCRs, along with other elements that emerged during the 1980s (refined automated on-site logistics, climbing robots, site cover technology, simulation and real-time monitoring technology, robot-oriented design), they are referred also as "integrated" automated construction sites. Almost all institutions involved in automation and robot technology from 1978 onwards switched their focus towards this new concept during the 1980s.

The reasons for this change in the general research and development direction were manifold. A major reason was that construction firms realized that STCRs that were not networked or embedded in a larger structured environment (SE) turned out to be incompatible with the way buildings were designed and built. STCRs substituted individual, mostly trades-related processes on-site. Common to all is that they were designed to execute certain tasks while activities of construction workers were not allowed to interfere significantly with the robots' activities. However, it turned out that under these premises, only very few robots could be used efficiently or economically. The constraints for the workers, the necessary safety regulations, coupled with the unforeseen, unpredictable, and dynamic processes at the site, set rigid boundaries for the individual robots working in parallel to the normal construction sequence.

Although STCRs allowed for high working speeds (and thus high productivity on a machine level), significant time had to be spent on-site for preparation, configuration, transportation, programming, and adjustment of spray volumes or injection pressures. Also, in most cases, the use of such systems was not considered in the design of the building (e.g., floor plans were not adjusted to the operation requirements of concrete flow finishing robots) or with the conventional job site installation (e.g., no parking or recharging stations for robots were foreseen or available). Furthermore, STCRs were not compatible with or integrated with upstream or downstream processes, and any productivity gains were equalized.

Apart from the aforementioned incompatibilities STCRs caused on (conventional) construction sites, the emergence of concepts for integrated sites was also nurtured by the new technological possibilities. Important for the integration of such systems into larger and coordinated automated systems was the development of systems that allowed the controlling and monitoring of an uninterrupted flow of information and material on-site between individual automated (and also nonautomated) entities that are involved in the (final) assembly of the building on-site. The emergence of concepts for integrated sites can also be explained from an evolutionary point of view. Most technological systems evolve from an array of single and disconnected entities to more complex systems, thanks to information and communication technology. Also, in many firms or industries, the movement from workshop-like production with individual and only loosely coupled production entities or stations to flow-line-like or production-line–like systems, with stable processes and continuous material flow was part of the evolution of those industries (e.g., automotive industry, computer industry).

From 1985 onwards, a working group set up by the Architectural Institute of Japan (from 1983) around Hasegawa and WASCOR (WASeda Construction Robot) group members, and joined by researchers of all major Japanese construction firms and equipment suppliers, intensified research in the development of site factories that would provide the SE required for automation and the efficient integration of STCRs and prefabricated components optimized for automated construction (see also **Volume 1**). In 1989, Shimizu tested the subsystems of the first integrated automated site on its research site with an experimental construction. In 1990 and 1992 the working group released several reports (e.g., in 1990: *Robotic Technology for Building Production* and in 1992: *Construction Automation Technology*) that provided the major Japanese construction companies with guidelines for the development and deployment of integrated, automated construction sites during the 1990s. In 1993 Shimizu started to use its so-called SMART (Shimizu Manufacturing System by Advanced Robotics Technology) commercially and all other major Japanese construction firms followed during the next decade.

3.1 Development and Refinement of Automated On-Site Logistics

On-site logistics can be considered as a kind of internal factory logistics (see also **Volume 1** for a detailed description of basic terminology). The development and refinement of automated on-site logistics systems was a key element of the step-by-step evolution of integrated automated construction sites. On-site logistics systems enabled the connection of individual work areas and work stations, and thus to advance from unorganized workshop-like activity (including STCR activity) to more organized (e.g., flow-line–like) manufacturing forms. The connection of workstations, work areas, machines, and STCRs by on-site logistic systems allows a factory-like controlled, continuous, and uninterrupted material flow. This can be seen as the basis for stable and productive "manufacturing" processes on the construction site.

Figure 3.1. Integration of Fujita subsystems and STCRs. (Visualization according to *Construction Robot System Catalogue in Japan*, 1999, p. 223.)

On-site logistics systems allow individual robots or groups of robots (e.g., ground-based factories) to be connected in terms of material flow with the positioning and assembly systems (e.g., sky factories [SF]).

On-site logistics systems allow the optimized utilization of advanced and costly high-tech equipment (STCRs, robotic overhead cranes, etc.) as they guarantee their continuous and uninterrupted operation by a supply of material. Within the analysis and discussion of STCRs, various exemplary, automated or robotized site logistics systems developed from 1980 onwards, and that formed the basis for high-performance logistics systems later deployed on automated/robotic on-site factories (focal point of **Volume 4**), were outlined. Fujita's automated facade panel installation system was the first deployed system in which refined logistics systems combining horizontal and vertical material transportation were used to directly link an on-site ground factory (on-site prefabrication of panels from parts and components) with the final assembly and positioning of the panel at the building's facade (Figure 3.1), and other concepts followed (Figure 3.2).

3.2 Development of Climbing Robots

Climbing mechanisms are an elementary technology present on all integrated automated construction sites. They allow the controlled or autonomous movement of the means of production (factory covers, equipment, working platforms) with the general working direction (e.g., vertical related to high-rise construction). From 1985 onwards, slip forming technology was refined, automated, and advanced towards systems that allowed automated climbing of working and equipment platforms. Besides the climbing mechanism itself, technology (software and hardware) to precisely control jacks, measure positions (e.g., of systems or factories that are moved by the

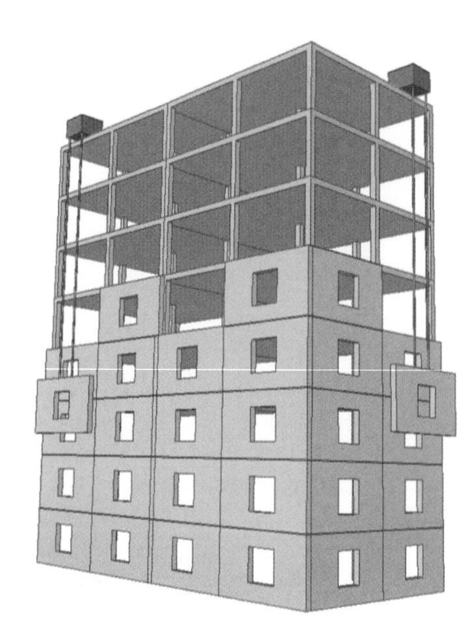

Figure 3.2. Concept for automatic facade lifting, positioning, and finishing by Japan Industrial Robot Association. (Visualization according to Hasegawa, 1999, p. 245.)

climbing mechanism), and for automated self-alignment of those systems and factories by the jack system was developed by researchers.

A special type of climbing system (CS) aimed – in contrast to other CSs – at raising or moving not the working platforms (see Figure 3.3), but the produced structure, so that the construction process takes place on the ground level. Such concepts follow the traditional way of building wooden buildings by first constructing the roof before pushing it up. A prototype for a modern version of such a system is Fujita's Arrow Up system (Figure 3.4). The principle of this construction system is to finish a steel structure in such a way that all assembly work can be done on the ground floor level. A stationary, temporary frame with press cylinders serves as a template around which the steel frame structure for each floor is erected and then pushed up. The system was designed to erect the steel structure of vertical parking garages that are typical for Japan. The advantages of such a push-up strategy were seen in improved working conditions and simplified material handling. For commercial use, it was necessary to be able to build over 10 floors.

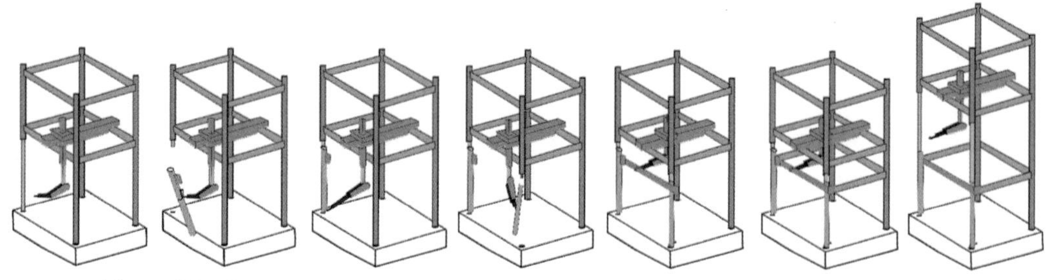

Figure 3.3. First concepts of a combination of a climbing mechanism with a robotic manipulator, allowing floor-by-floor installation of columns and beams. (Visualization according to Obayashi, H. Teraoku.)

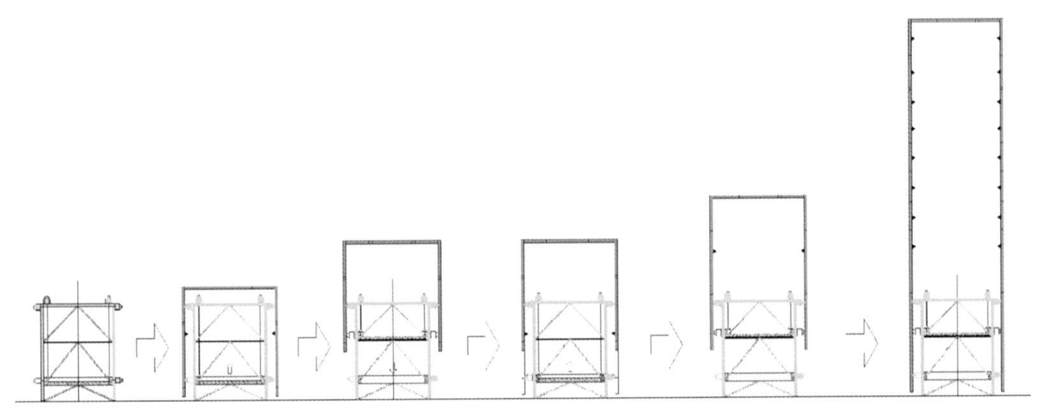

Figure 3.4. Fujita's Arrow Up allowed the subsequent push-up of a steel structure assembled on a ground floor level. (Visualization according to *Construction Robot System Catalogue in Japan*, 1999)

3.3 Refinement of Site-Cover Technology

Site covers ensure a work space is unaffected by the weather. Furthermore, they allow for a workspace to be illuminated properly (e.g., at night, for example) and ensure that the noise of work activity does not disturb the neighborhood (e.g., at night, for example). The cover frames can – if dimensioned properly – provide a carrying frame for (automated) logistic systems, overhead cranes, and positioning systems. Site covers reduce the number of unforeseen events (and thus negative impacts) on the organization on-site. They are the basis for creating a factory-like SE on-site. Both in Russia and in Germany, from the 1930s onwards, sites were sometimes covered to allow the use of the workforce during the entire day (24 hours) in several shifts, over the whole week (24/7), and during the whole year (to enable both summer and winter construction). In prewar and during war time in particular, this approach was supposed to increase speed and efficiency of strategically important construction projects. Construction-like site-covers with integrated logistics and crane systems were also developed from that time onwards both in aircraft manufacturing and ship building (for further information, see **Volume 1**). During the 1980s, in Japanese civil engineering projects, such as in the construction of power plants or infrastructure projects, site covers were increasingly deployed as systems integrated with and thus were refined by automated logistics, automated overhead cranes, and other automated machines (Figure 3.5).

3.4 Introduction of Simulation and Real-Time Monitoring Technology in Construction

During the 1980s, major Japanese LSP companies introduced enterprise resource management systems (e.g., Sekisui Heim's Heim automated parts pickup system [HAPPS]; see also **Volume 2**) into construction that were able to generate material lists (bill of materials) as well as work procedures, logistics operations, and information for machines. At the same time, manufacturing industries relying on automation, such as the automotive industry, started to develop software tools that allowed them to simulate the production process (body in white assembly) in advance. From 1985

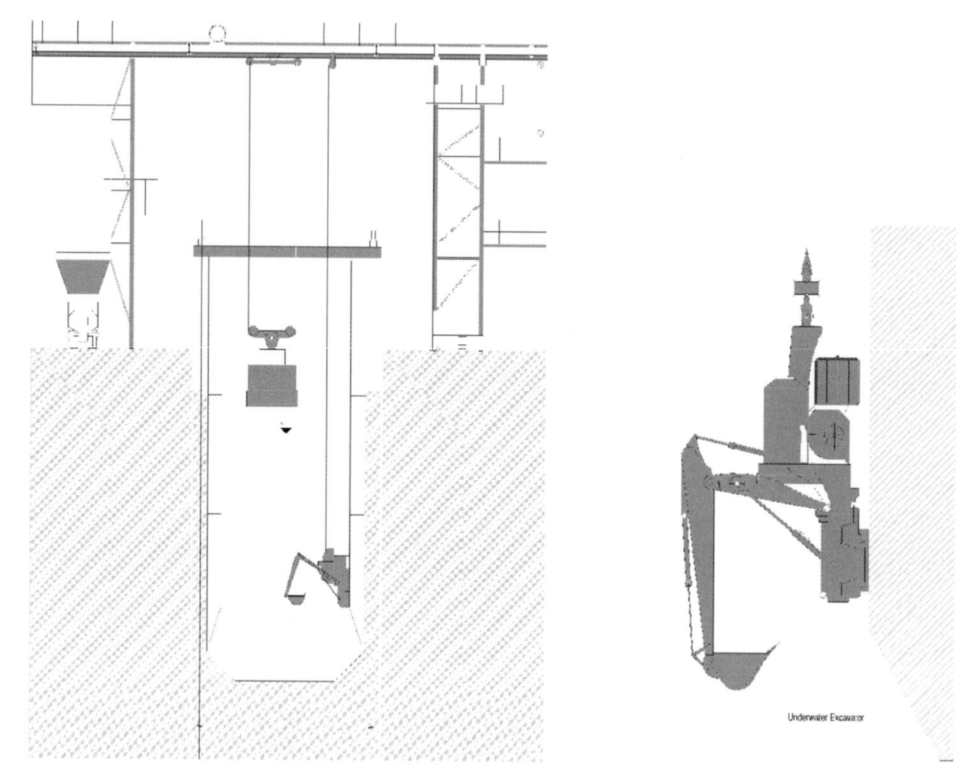

Figure 3.5. Super open caisson system (SOCS) used and taken further by a multitude of construction companies from the end of the 1980s onwards. The spanning and covering of on-site work by a gantry-type site factory, integrating multiple automated subsystems, is shown. (Visualization according to *Construction Robot System Catalogue in Japan*, 1999.)

onwards, research activity increasingly aimed at using simulation, enterprise resource planning (ERP), and manufacturing process-monitoring technology not only off-site in controlled factory environments, but also on the construction site. For the evolution of integrated, automated construction sites, the development of the aforementioned simulation, control, and monitoring tools was just as important as the development of the machines and robots themselves, as automation on-site demanded:

1. That the construction process (developing production/manufacturing process) be planned and scheduled in detail.
2. That the deployment of complex and costly robot systems could be simulated and optimized.
3. That the material flow should not only be executed physically by a logistic system, but should also be controlled on an informational level to achieve just in time (JIT)/just in sequence (JIS) delivery of exactly the demanded components on-site.

At WASCOR, the central institute directing the construction automation attempts during the 1980s in Japan, computer-aided engineering systems and computer-aided management systems were seen as complementary to and necessary in the deployment of construction automation. Since the 1990s, Shimzu, as well as other companies, have been developing simulation tools that allow the simulation and optimization of construction processes on-site. Most companies that deployed automated construction sites used such tools to optimize the configuration of the automated/robotic on-site factory (for an examination of those tools, see analysis of subsystems of automated/robotic on-site factories in **Volume 4**). Shimzu even developed a simulator that allowed the simulation and optimization of the disassembly of the site factory after completion of the building (see **Volume 4**).

3.5 Introduction of Robot-Oriented Design Strategies in Construction

From 1985, in light of the continuous deployment of more and more robots and automation technology on construction sites in Japan, various institutions (researchers, companies, and universities) started to adapt the design of buildings (structure, modules, components, joining systems, etc.) to support and simplify the use of robots on-site, and thus to enhance their efficiency. Approaches that aim at redesigning the product to allow for more efficient processes and the efficient use of humans and machines in manufacturing were pioneered before 1985 by Toyota Motors. The Toyota Production System (TPS) applied strategies that were aimed at optimizing designs for failure free assembly, JIT logistics, and processing by automated systems and robots (for further details, see **Volume 1**). Furthermore, traditional Japanese timber construction, as well as modern Japanese LSP (starting in the 1970s), relied on systemized building designs that simplified and reduced on-site construction efforts (for further information, see **Volume 1**). From 1985, a working group built around Hasegawa and the WASCOR group developed strategies for changing building design to support the operation of automated systems and robots on the construction site.

The basis for the development of WASCOR's strategies was the research activity (and later on doctoral thesis) of Bock (1989). He developed the concept of robot-oriented design (ROD) at the University of Tokyo under the supervision of Professor Uchida (who also a decade before supervised the development of Sekisui Heim's M1, the first building kit designed for production line–based mass production). ROD aimed at the reduction of complexity of the assembly process by the reduction of parts and the redesign of component structures as a prerequisite for the application of automation and robotics on the construction site (Figure 3.6). In parallel, researchers started to develop construction-specific kinematics science and to identify and specify the degrees of freedom (DOF), robot motions, and robot trajectories necessary to install specific parts and components on the construction site (Figure 3.7).

3.6 First Concepts for Integrated Sites: Cooperating STCRs

After several hundred STCRs had already been deployed, concepts evolved during the 1980s that envisioned various STCRs working in parallel, cooperatively, and remotely controlled on the construction site. The concepts of tele-existence and tele-control of multiple construction machines were pioneered by Professor Susumu Tachi at the University of Tokyo in the 1980s. Applications for tele-operated

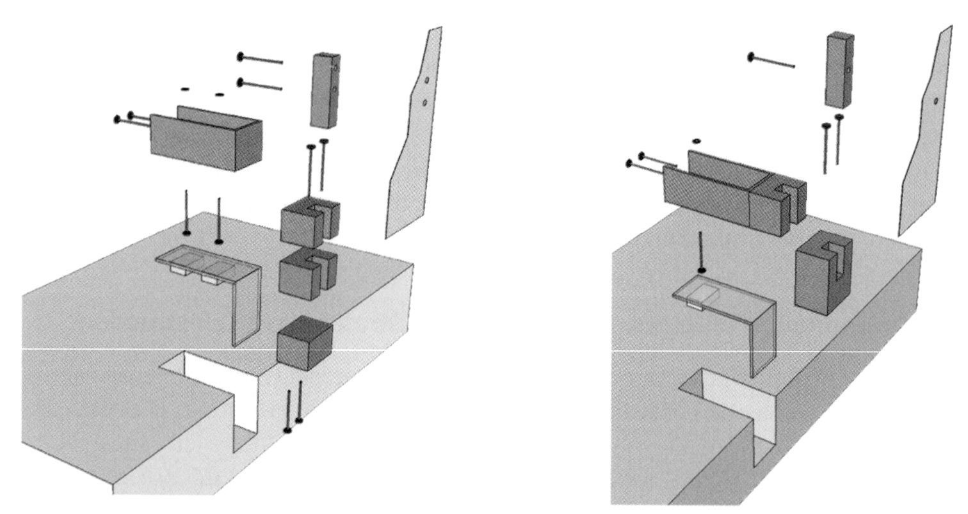

Figure 3.6. Development of method of ROD by T. Bock from 1985 onwards within WASCOR. Reduction of complexity of assembly process by reduction of parts and redesign of component structure. (Visualization according to Bock 1988.)

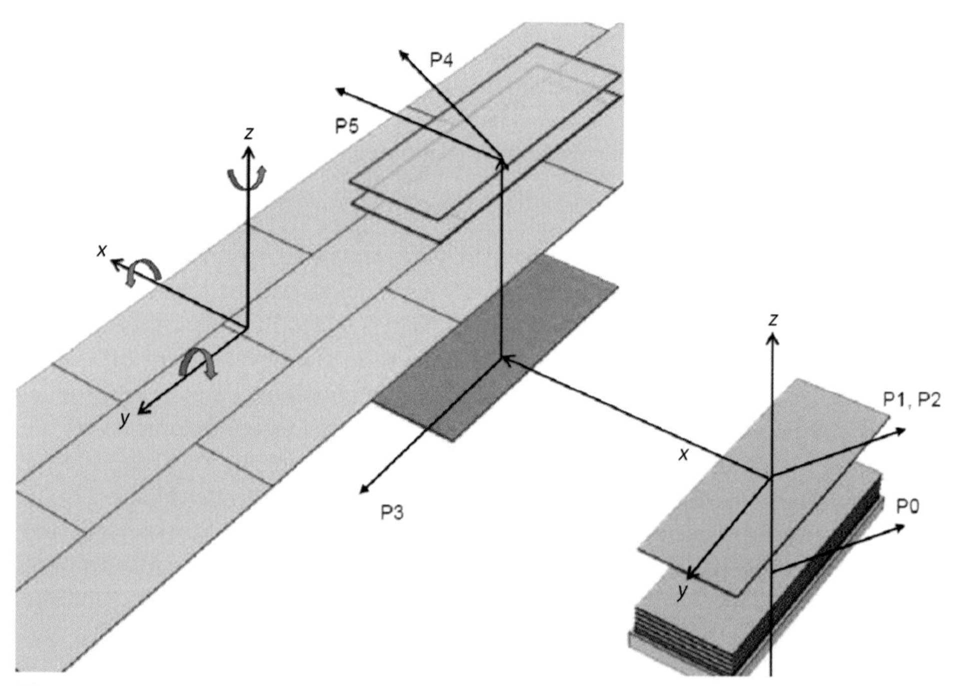

Figure 3.7. Analysis of positioning process and related kinematics of the installation of a ceiling panel as a basis for automation. Researchers started to develop a construction-specific kinematics science and to identify and specify DOF, robot motions, and robot trajectories necessary to install specific parts and components on the construction site. (Image adopted from Hasegawa et al. 1992.) P0, start position; P1, picking up panel; P2, rotating panel; P3, delivering panel; P4, lifting up panel; P4, positioning of panel. (Visualization according to Hasegawa et al. 1992.)

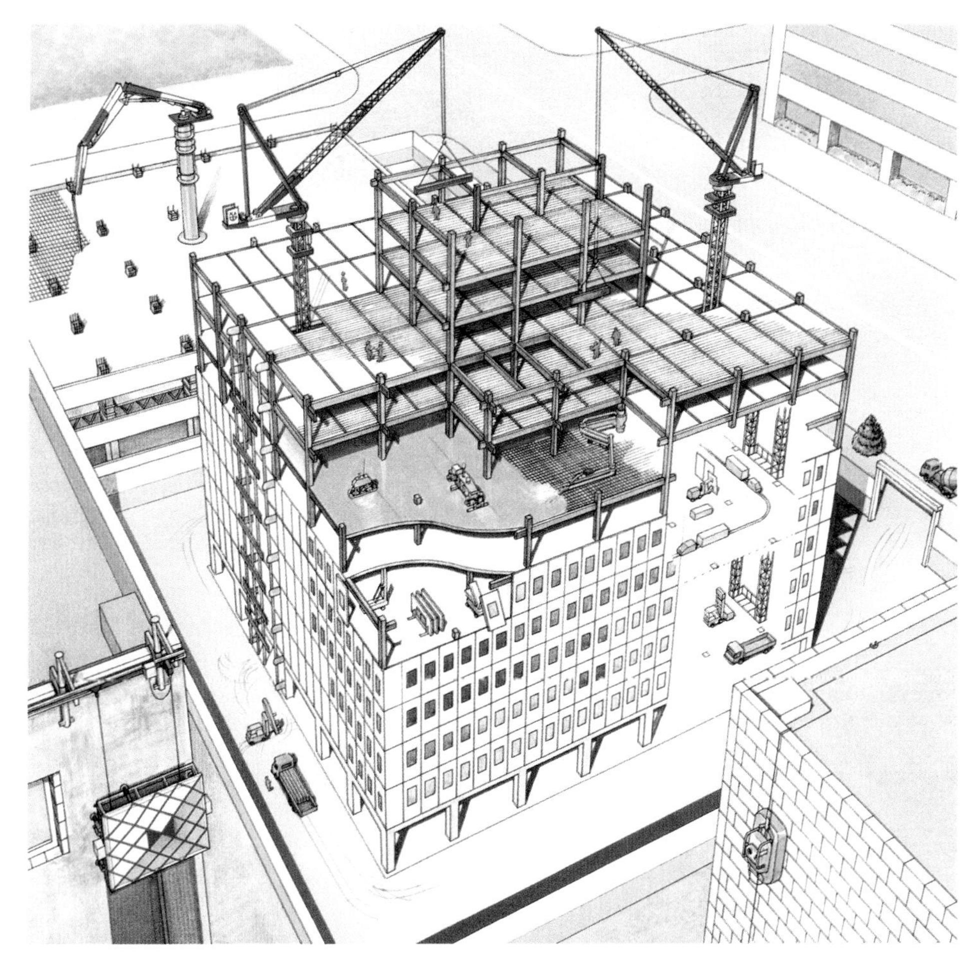

Figure 3.8. Integration and cooperation of several robotic systems on the construction site to construct a building on-site. Concept discussed within WASCOR, Takenaka Corporation. (Image: Takenaka Corporation.)

construction machines (excavators and trucks) had been used in Japan in the real world, for example, during the Mount Unzen disaster incident, where a volcanic eruption covered a large area with dust and remotely controlled robotic construction equipment was used to clean the area. Concepts and knowledge about tele-operation and the remote control of groups of construction equipment inspired concepts for the control of multiple cooperating construction robots on the building construction site (Figure 3.8).

3.7 First Concepts for Integrated Sites: Factory Approach

In the second half of the 1980s more and more concepts evolved that analysed the possibility of installing a structured workplace or moving factory on construction sites. In contrast to the concept of networked, tele-operated, and cooperating STCRs (see Section 3.6), those approaches fused assembly and logistics systems more with

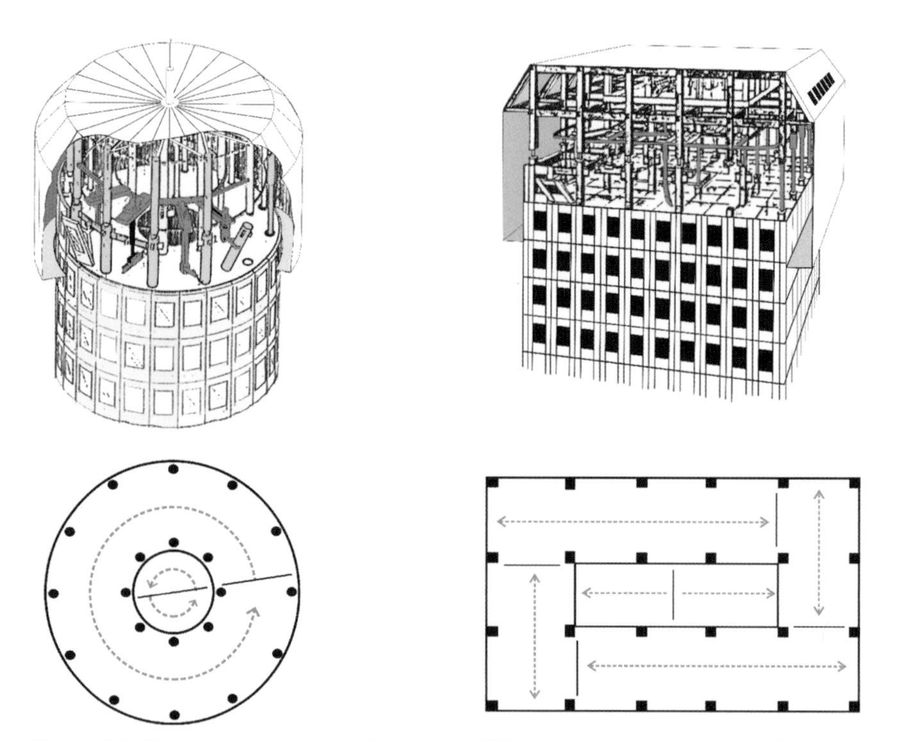

Figure 3.9. First concept studies for on-site SF and arrangement of overhead manipulator system automatic building construction floor. (Image: Visualization according to Obayashi Research Institute/Teraoku.)

Figure 3.10. First concept studies for on-site SF sitting on top of the building by Obayashi. The round design of the building would be in tune with an efficient operation of the robotic overhead manipulators and could unfortunately be applied only as a rectangular system. (Image: Obayashi.)

the covering frame structures and focused, in contrast to the concept of cooperating STCRs (used primarily for facade installation and interior work), more strongly on the erection of the building's load-bearing main structure (Figures 3.9 and 3.10). Later, the first commercially deployed integrated, automated construction sites fused both approaches.

4 From Stand-Alone Solutions to Systems Integrated by Structured Environments

The Shimizu Corporation was the first Japanese construction company to transfer its experience of more than 12 years of construction robot development to the concept of a fully automated, integrated building construction site. Already in 1989, at the research site of the construction company Shimizu Corporation, subsystems of the SMART (Shimizu Manufacturing system by Advanced Robotics Technology) climbing mechanism, overhead assembly robots, and logistics system were tested with the construction of an experimental building.

Approximately five years of development and a financial outlay of more than €12 million were necessary to operate prototype sites with SMART in 1990 and 1991. SMART integrated a moving, combined logistics and assembly system in a sky factory for steel structure erection with single-task construction robot (STCR) technology for interior finishing. The SMART system was used commercially for the first time in 1991 for the construction of the 20-storey Juroku Bank in Nagoya (for further details on this system, see analysis in **Volume 4**). In the experimental construction site, the automated construction process was limited to the simulation of the positioning and welding of a building's steel skeleton, the positioning of precast concrete floor elements, and the positioning of exterior and interior wall elements. A digital management system allowed real-time control of the site, the automated logistics (vertical and horizontal distribution of components such as, e.g., columns), and the automated welding process.

All in all, it can be concluded that development of automated/robotic on-site factories was based on the insight from 1985 on that the newly developed machine technology in the form of STCRs was not able to provide a solution for the improvement of construction processes. STCRs were too task specific and conflicted with conventional construction work on site. Furthermore, aspects of continuous operation and material supply could not be addressed by them. STCRs addressed only subproblems of the construction process, but did not address the main challenge in construction, which is to build up the building's main structure (in this book series also referred to as "frame" or "platform", for more details on frame and infill and platform strategies, see **Volume 1**) on the construction site – along a main working direction – and to equip it with technical infill and with finishing. The erection of a building can be considered as a complex assembly process. As in any other industry, the systematization, mechanization, and finally automation of complex manufacturing processes can be achieved only by the cooperation of multiple machine systems with other means of production in a structured environment (SE).

STCRs, from a technological point of view and with respect to the improvement of productivity related to specific construction tasks, represent an enormous improvement. However, from a larger organizational point of view that considers the whole assembly process on-site, the application and integration of STCRs into the construction process could still be considered as workshop-like production with individual and only loosely coupled production entities or stations. After the evaluation of the first series of STCRs a change in direction was made towards the integration of STCRs to flow-line–like or production-line–like systems, with stable processes and continuous material flow. The integration of multiple machine systems with other means of production necessitated the set-up of SEs on the construction sites SMART Prototype on Shimzu's research testing area in 1989. During the 1990s all major Japanese contractors and machine builders followed this approach and integrated their STCR technology together with other new technologies and approaches within automated/robotic on-site factories. Individual companies in Europe and Korea followed.

References

Andreas Nüchter; Jan Elseberg; Dorit Borrmann (2013) "Irma3D – An intelligent robot for mapping applications". In *Proceedings of the 3 IFAC Symposium on Telematics Applications (TA '13)*, Vol. 3, Part 1, Seoul, Korea, November 2013, doi:10.3182/20131111-3-KR-2043.00011.

Andres, J.; Bock, T.; Gebhardt, F.; Steck, W. (1994) "First results of the development of the mansonry robot system ROCCO: A fault tolerant assembly tool". In *Proceedings of the 11th International Symposium on Automation and Robotics in Construction*, pp. 87–93. Amsterdam: International Association for Automation and Robotics in Construction.

Aoyagi, H.; Shibata, Y. (1988) "Development of the horizontal distributor for concrete placing". In *Proceedings of the 5th International Symposium on Automation and Robotics in Construction*, pp. 541–550. Tokyo: International Association for Automation and Robotics in Construction.

Arai, M.; Hoshino, H.; Sugata, M.; Tazawa, S.; Hayashida, H. (2010) "Development of closed type robot for removing and recollection of sprayed asbestos materials on steel beams". In *Proceedings of the 12th Symposium on Construction Robotics in Japan*, pp. 289–294. Tokyo: Japan Industrial Robot Association (JIRA).

Architectural Institute of Japan (AIJ). (1988) *Proceedings Ascending Technologies* http://www.asctec.de/ (Accessed 31 December 2015).

Architectural Institute of Japan (AIJ) (1990) *Proceedings of the 1st Symposium on Construction Robotics in Japan.*

Architectural Institute of Japan (AIJ) (1998) *Proceedings of the 7th Symposium on Construction Robotics in Japan.*

Architectural Institute of Japan (AIJ) (2008) *Proceedings of the 11th Symposium on Construction Robotics in Japan.*

Architectural Institute of Japan, Committee of Construction Material, Subcommittee of Construction Robotics Technology (1989) *Proceedings of the 3rd Construction Robot Symposium.*

Architectural Institute of Japan, Committee of Construction Material, Subcommittee of Construction Robotics Technology (1993) *Proceedings of the 7th Construction Robot Symposium.*

Architectural Institute of Japan, Research Committee on Building Materials and Construction Procedure, Subcommittee on Robotics Technology in Building Construction (1988) *Proceedings of the 2nd Construction Robot Symposium*. Tokyo.

Architectural Institute of Japan, Research Committee on Building Materials and Construction Procedure, Subcommittee on Automation in Building Construction (1995) *Proceedings of the 9th Construction Robot Symposium*. Tokyo.

Architectural Institute of Japan, Research Committee on Building Materials and Construction Procedure, Subcommittee on Automation in Building Construction (1996) *Proceedings of the 10th Construction Robot Symposium*. Tokyo.

Atsuhiro, D.; Koji, H.; Tomoya, K.; Takashi, S. (2002) "Development and application of automated delivery system for finishing building materials". In *Proceedings of 19th International Symposium on Automation and Robotics in Construction 2002*, pp. 235–241. Washington, DC: International Association for Automation and Robotics in Construction.

Autodesk (2015) Architecture and construction software. http://www.businesswire.com/multimedia/home/20150430006461/en/#.VVibfJPpXU8 (Accessed 17 May 2015).

Automated Building Construction System (1993) Riverside Sumida Bachelor Dormitory documentary video [VHS]. Tokyo: Obayashi Corporation.

Automation in Construction, An International Research Journal, Vol. 12 Number 2 March 2003, 12(2) 113–226 (2003).

Automation in Construction, An International Research Journal, Vol. 12 Number 4 July 2003, 12(4) 349–464 (2003).

Autonomous Solutions, Inc. (ASI) http://www.asirobots.com/ (Accessed 31 December 2015).

Bennett, J.; Flanagan, R.; Norman, G. (1987) *Capital & Countries Report: Japanese Construction Industry*. Reading, UK: Centre for Strategic Studies in Construction.

Bock, T. (1989) *Robot Oriented Design*. Dr.-Ing. dissertation, Faculty of Engineering, Chair of Building Production, the University of Tokyo.

Bock, T. (1998) "Robotics in Architecture". In *Proceedings of La Science et la Technologie au Service de l'Architecture, Conference Europ enne sur l'Architecture UNESCO*. Paris: Center Robotique Intégrée Ile de France.

Bock, T.; Gebhart, F. (1994) "ROCCO-Robot Assembly System for Computer Integrated Construction, An Overview", In: *First European Conference on Product and Process Modelling in the Building Industry*, pp. 583–591. Dresden, Germany.

Bock, T.; Steffani, H. F. (1994) "A Semi Autonomous Vehicle for a Mobile Robot (ROCCO)", In: *First European Conference on Product and Process Modelling in the Building Industry*, pp. 591–599. Dresden, Germany.

Bock, T.; Linner, T. (2010b) "From early trials to advanced computer integrated prefabrication of brickwork". In G. Girmscheid; F. Scheublin (eds.), *New Perspective in Industrialization in Construction – A State-of-the-Art Report*. CIB Publication 329, pp. 161–181. Zürich: ETH Zürich.

Bock, T.; Linner, T.; Eibisch, N.; Lauer, W. (2010) "Fusion of product and automated-replicative production in construction". In *27th International Symposium on Automation and Robotics in Construction*, pp. 12–21. Bratislava: International Association for Automation and Robotics in Construction.

Bock, T.; Linner, T.; Georgoulas, C.; Mayr, M.; Meyer-Andreaus, J. (2012) "Innovation deployment strategies in construction". In *Proceedings of Creative Construction Conference*, pp. 105–118. Budapest, Hungary.

Borrmann, D.; Afzal, H.; Elseberg, J.; Nüchter, A. (2012) "Mutual calibration for 3D thermal mapping". In *Proceedings of the 10th Symposium on Robot Control (SYROCO)*, pp. 605–610. Amsterdam: Elsevier.

Bosscher, P., Williams II, R. L.; Bryson, L. Sebastian; Castro-Lacouture, Daniel (2007) "Cable-suspended robotic contour crafting system". *Automation in Construction*, 17: 45–55.

Bräunl, T. (2008) *Embedded Robotics, Mobile Robot Design and Application with Embedded Systems*, 3rd ed. Berlin and Heidelberg: Springer-Verlag.

Bridgit (2015) Construction software. http://gobridgit.com (Accessed 17 May 2015).

Brown, S. F. (1988) "Meet the Wallbots". *Popular Science*, August: 54–55.

Brunkenberg Systems AB http://www.brunkeberg.com/ (Accessed 31 December 2015).

Caldas, C. H.; O'Brien, W. J. (2009) "Computing in civil engineering". In *Proceedings of the 2009 international Workshop on Computing in Civil Engineering*. Austin, TX: American Society of Civil Engineers.

Cheng, T.; Teizer, J. (2011) "Crane operator visibility of ground operations". In *Proceedings of 28th International Symposium on Automation and Robotics in Construction*, pp. 699–705. Seoul, Korea: International Association for Automation and Robotics in Construction.

Cho, C.-Y.; Kwon, S.-W.; Lee, J.; You, S.-J.; Chin, S.-Y.; Kim, Y.-S. (2009) "Basic study of smart robotic construction lift for increasing resource lifting efficiency in high-rise building construction". In *Proceedings of 26th International Symposium on Automation and Robotics*

in Construction, pp. 266–277. Austin, TX: International Association for Automation and Robotics in Construction.
Clearpath Robotics, Inc. http://www.clearpathrobotics.com/ (accessed 31 December 2015).
Construction Robotics http://www.construction-robotics.com/ (Accessed 31 December 2015).
Council for Construction Robot Research (1999) *Construction Robot System Catalogue in Japan*. [research report]. Tokyo: Japan Robot Association.
Cousineau, L.; Miura, N. (1998) *Construction Robots: The Search for New Building Technology in Japan*. Reston: American Society of Civil Engineers (ASCE).
Cyberdyne Inc. http://www.cyberdyne.jp/english/ (Accessed 31 December 2015).
Dalacker, M. (1997) *Entwurf und Erprobung eines mobilen Robotters zur automatisierten Erstellung von Mauerwerk auf der Baustelle*. Schriftenreihe Planung, Technologie Management und Automatisierung im Bauwesen. Munich: Fraunhofer IRB Verlag.
Daewoo Shipbuilding and Marine Engineering (DSME) (2015) http://www.dsme.co.kr (Accessed 17 May 2015).
Daftry, S.; Hoppe, C.; Bischof, H. (2015) "Building with drones: Accurate 3D façade reconstruction using MAVs". In *IEEE International Conference on Robotics and Automation (ICRA '15)*, Seattle, WA.
Doggett, W. (2002) "Robotic assembly of truss structures for space systems and future research plans". In *Aerospace Conference Proceedings*, 2002, Vol. 7, pp. 3589–3598. Big Sky, MT: IEEE.
Doka GmbH https://www.doka.com (Accessed 31 December 2015).
Dong, X.; Yurong M.; Haile E. (2005) "Work-related fatal and non-fatal injuries among construction workers: 1992 2003". Silver Spring, MD: The Center to Protect Workers' Rights (CPWR).
Ebihara, M. (1988) "The development of an apparatus for diagnosing the interior condition of walls". In *Proceedings of 5th International Symposium on Automation and Robotics in Construction*, 867–876. Tokyo: International Association for Automation and Robotics in Construction
Elkmann, Norbert; Felsch, Torsten; Forster, Tilo (2010) "Robot for rotor blade inspection". In *2010 1st International Conference on Applied Robotics for the Power Industry (CARPI)*, Vol. 1, pp. 1–5, 5–7 October 2010, doi: 10.1109/CARPI.2010.5624444.
Elkmann, Norbert; Felsch, Torsten; Sack, M.; Saenz, José; Hortig, Justus (2002) "Innovative service robot systems for facade cleaning of difficult-to-access areas". In *IEEE/RSJ International Conference on Intelligent Robots and Systems 2002. Proceedings*, Vol. 1, 30.9.–4.10.2002, EPFL Lausanne, Switzerland. Piscataway, NJ: IEEE Operations Center.
EZ Renda http://www.ezrenda.com/ (Accessed 31 December 2015).
Fastbrick Robotics http://www.fbr.com.au/ (Accessed 31 December 2015).
Feldmann, K.; Koch, M. (1998) "A mobile robot system for assembly operations at interior finishing". In *Proceedings of 15th International Symposium on Automation and Robotics in Construction*, pp. 93–102. Munich, Germany: International Association for Automation and Robotics in Construction.
Ferguson, E. S. (1962) "Kinematics of mechanisms from the time of watt". In *Contributions from the Museum of History and Technology Washington*, Paper 27, pp. 185–230.
Flight Assembled Architecture – Joint Project by Raffaello D'Andrea and Grammazio and Kohler [http://raffaello.name/dynamic-works/flight-assembled-architecture/ (Accessed 21 January 2014).
Future Cities Laboratory, Design of Robotic Fabricated High Rises, Prof. Fabio Gramazio, Prof. Matthias Kohler, Future Cities Laboratory of the Singapore-ETH Centre for Global Environmental Sustainability, http://www.fcl.ethz.ch/module/digital-fabrication/ (Accessed 31 December 2015).
Gambao, E.; Hernando, M. (2006) "Control system for a semi-automatic facade cleaning robot". In *Proceedings of 23rd International Symposium on Automation and Robotics in Construction*, pp. 406–411. Tokyo: International Association for Automation and Robotics in Construction.
Gambao, Ernesto; Hernando, Miguel; Surdilovic, Dragoljub (2008) "Development of a semi-automated cost-effective facade cleaning system". In Carlos Balaguer; Mohamed

Abderrahim (eds.), *Robotics and Automation in Construction*. http://www.intechopen.com/books/robotics_and_automation_in_construction/development_of_a_semi-automated_cost-_effective_facade_cleaning_system.

General Robotics, Automation, Sensing and Perception (GRASP) Laboratory, Penn Engineering, https://www.grasp.upenn.edu/research-groups/kumar-lab (Accessed 31 December 2015).

General Robotics, Automation, Sensing and Perception (GRASP) Laboratory, Penn Engineering, www.grasp.upenn.edu (Accessed 21 January 2014).

GGR Group http://www.ggrgroup.com/ (Accessed 31 December 2015).

Gogget, W. (2002) "Robotic assembly of truss structures for space systems and future research plans". Hampton, VA: NASA Langley Research Center.

Hasegawa, Y. (1985) "Robotization of reinforced concrete building construction in Japan". In *Proceedings of 3rd International Symposium on Automation and Robotics in Construction*, pp. 423–433. Marseille, France: International Association for Automation and Robotics in Construction.

Hasegawa, Y. (1999) *Robotization in Construction*. Tokyo: Kogyo Chosakai.

Hasegawa, Y.; Onoda, T.; Tamaki, K. (1992) "Modular robotic application for flexible building construction". In *Proceedings of 13th International Symposium on Automation and Robotics in Construction*, pp. 113–122. Tokyo: Japan International Association for Automation and Robotics in Construction.

Hjelle, David Alan; Lipson, Hod (2009) "A robotically reconfigurable truss". In *Proceedings of ASME/IFToMM International Conference on Reconfigurable Mechanisms and Robots (ReMAR 2009)*, pp. 73–78. London: King's College London.

Honda Motor/Honda Robotics Division http://world.honda.com/HondaRobotics/index.html (Accessed 31 December 2015).

Hortig, J.; Elkmann, N.; Felsch, T.; Sack, M.; Böhme, T. (2001) "The Sirius robot platform for tool handling on large vertical surfaces". In *Proceedings of International Symposium on Automation and Robotics in Construction 2001*, pp. 1–4. Krakow, Poland: International Association for Automation and Robotics in Construction.

IEEE Spectrum (2015) "Korean shipbuilder testing industrial exoskeletons for future cybernetic workforce. http://spectrum.ieee.org/automaton/robotics/industrial-robots/korean-shipbuilder-testing-industrial-exoskeletons-for-future-cybernetic-workforce (Accessed 17 May 2015).

Institute for Advanced Architecture of Catalonia (IAAC) robots.iaac.net (Accessed 31 December 2015).

Inoue, F.; Doi, S.; Okada, T.; Ohta, Y. (2009) "Development of automated inspection robot and diagnosis method for tile wall separation by wavelet analysis". In *Proceedings of International Symposium on Automation and Robotics in Construction 2009*, pp. 379–388. Austin, TX: International Association for Automation and Robotics in Construction.

International Association for Automation and Robotics in Construction (IAARC) (1998) *Robots and Automated Machines in Construction*. Watford, UK: IAARC.

Itou, M.; Nishita, K.; Houkyou, M. (1994) "Application of all-weather automatic building construction systems". In *Construction Robot Symposium*. Tokyo.

Iturralde, K.; Linner, T.; Bock, T. (2015–1) "Comparison of automated and robotic support bodies for building facade upgrading". In *Proceedings of the 32nd International Symposium on Automation and Robotics in Construction and Mining (ISARC2015)*, Oulu, Finland, 15–18 June, 2015, pp. 1–8.

Iturralde, K.; Linner, T.; Bock, T. (2015-2) "Development and preliminary evaluation of a concept for a modular end-effector for automated/robotic facade panel installation in building renovation". In *10th Conference on Advanced Building Skins*, 4–5 October 2015, Bern, Switzerland, pp. 4662–4671.

Jackson, J. R. (1990) *Robotics in the Construction Industry*. Master's thesis, University of Florida.

Japan Federation of Construction Contractors (2012) *Construction Industry Handbook 2012* [research report]. Tokyo: Japan Federation of Construction Contractors.

Japan Industrial Robot Association (JIRA) (1988) *Handbook of the Industrial Robot Introduction – Non-manufacturing Industry 1988*. Tokyo: JIRA.
Japan Industrial Robot Association (JIRA) (1990) *The Specifications and Applications of Industrial Robots in Japan – No-manufacturing Fields*. Tokyo: JIRA.
Japan Industrial Robot Association (JIRA) (1991) *Proceedings of the 2nd Symposium on Construction Robotics*. Tokyo: JIRA.
Japan Industrial Robot Association (JIRA) (1993) *Proceedings of the 3rd Symposium on Construction Robotics*. Tokyo: JIRA.
Kajima (2000) AMURAD Documentary Video [VHS]. Tokyo: Kajima.
Kajima (2011) Interview with Kajima R&D staff at ISARC, Seoul, Korea, June.
Kajima Company Brochure "Automation technology for maintenance and reform". Tokyo: Kajima.
Kajima Company Brochure. "Automation technology for the construction of nuclear power plant". Tokyo: Kajima.
Kajima Company Brochure. "Construction robot/automation technology". Tokyo: Kajima.
Kameda, M.; Suzuki, A.; Watanabe, J., Nishigami, M. (1996) "Study on development and utilization of construction robots". In *Proceedings of the 13th International Symposium on Automation and Robotics in Construction*, pp. 141–150. Tokyo: International Association for Automation and Robotics in Construction.
Kangari, R.; Yoshida, T. (1989) "Prototype robotics in the construction industry". *Journal of Construction Engineering and Management*, 115(2): 284–301.
Kawada Robotics Corporation http://www.kawadarobot.co.jp (Accessed 31 December 2015).
Khoshnevis, B. (2003) "Automated construction by contour crafting – related robotics and information technologies". *Automation in Construction*, 13(2004): 5–19.
Kodaki, K.; Nakano, M.; Nanasawa, T.; Maeda, S.; Miyazawa, M. (1996) "Development of the automatic system for pneumatic cassion". In *Proceedings of International Symposium on Automation and Robotics in Construction 1996*, pp. 333–341. Tokyo: International Association for Automation and Robotics in Construction.
Komatsu http://www.komatsu.com/ (Accessed 31 December 2015).
Kumita, Y.; Takimoto, T.; Nozue, A.; Murakoshi, K. (1992) "Development of a desk/chair arrangement robot". In *Proceedings of the 9th International Symposium on Automation and Robotics in Construction*, pp. 525–534. Tokyo: International Association for Automation and Robotics in Construction.
Lauer, W.; Bock, T. (2010) "Location orientation manipulator by Konrad Wachsmann, John Bollinger and Xavier Mendoza". In *27th International Symposium on Automation and Robotics in Construction*, pp. 704–712. Bratislava: International Association for Automation and Robotics in Construction.
Lee, S.-Y. (2008) *Human-Robot Cooperation for Installing Bulk Building Materials at Construction Sites*. Dr.-Ing. dissertation, Department of Mechanical Engineering, Hanyang University, Seoul.
Lee, S. Y.; Gil, M.; Lee, K.; Lee, S.; Han, C. (2007) "Design of a ceiling glass installation robot". In *Proceedings of the 24th International Symposium on Automation and Robotics in Construction*, pp. 247–252. Madras: International Association for Automation and Robotics in Construction.
Lichtenberg, J.; Segers, S. (2000) "Mechanisation of ceramic tiling". [Research Report], End Report EC Contract No. BRST-CT98-5238, Project No. BES2-2676.
Lichtenberg, J. J. N. (2003) "The development of a robot for paving floors with ceramic tiles". In *Proceedings of the 20th International Symposium on Automation and Robotics in Construction*, pp. 85–88. Eindhoven, the Netherlands: International Association for Automation and Robotics in Construction.
LiDAR Equipped Robot, Department of Geological Engineering. Missouri University of Science and Technology. Principal investigators: Prof. Dr. Maerz, Prof. Dr. Ye Duan (2010).
Lindsey, Q; Mellinger, D; Kumar, V. "Construction with quadrotor teams". *Autonomous Robots*, 33(3): 323–336.

Linner, T. (2013) *Automated and Robotic Construction: Integrated Automated Construction Sites*. Dr.-Ing. dissertation, Technische Universität München. In particular chapters 1, 3 and 4 in this volume (**Volume 3**) are based on chapter 4 of the thesis; furthermore the classification system for the STCRs has been adopted from the thesis and expanded and detailed in this volume (**Volume 3**).

Llorens Bonilla, Baldin; Asada, H. Harry. (2014) "A robot on the shoulder: Coordinated human-wearable robot control using coloured petri nets and partial Least Squares predictions". In *2014 IEEE International Conference on Robotics and Automation (ICRA)*, pp. 119–125. Hong Kong.

Lobo, Daniel; Hjelle, David Alan; Lipson, Hod (2009) "Reconfiguration algorithms for robotically manipulatable structures". In *Proceedings of ASME/IFToMM International Conference on Reconfigurable Mechanisms and Robots (ReMAR 2009)*, pp. 13–22. London: King's College London.

Lockheed Martin (2015) "FORTIS exoskeleton". www.lockheedmartin.com/us/products/exoskeleton/FORTIS.html (Accessed 17 May 2015).

LOMAR s.r.l. http://www.infolomar.com/ (Accessed 31 December 2015).

Maeda, J. (1994) "Development and application of the SMART system". *Automation and Robotics in Construction*, 6: 457–464.

MAI International GmbH http://www.mai.at/ (Accessed 31 December 2015).

Matsuike, T.; Sawa, Y.; Ohashi, A.; Shinji Sotozono, Fukagawa, R.; Muro, T. (1996) "Development of automatic system for diaphragm-wall excavator". In *Proceedings of 13th International Symposium on Automation and Robotics in Construction 1996*, pp. 285–294. Tokyo: International Association for Automation and Robotics in Construction.

Miyazaki, Y., et al. "Development of 'multiple purpose construction hand' – multiple-purpose construction robot to mount exterior wall panels". Tokyo: Kajima Corporation & Komatsu Ltd.

dMorita, M.; Muro, E.; Kanaiwa, T.; Nishimura, H. (1993) "Study on simulation of roof pushup construction method". *Automation and Robotics in Construction*, 5: 1–8.

Nerdinger, W. (2010) "Die Zukunft aus der Fabrik – Konrad Wachsmanns 'Wendepunkt' im Kontext". In Wendepunkte im Bauen: von der seriellen zur digitalen Architektur, pp. 10–17. Munich: Nerdinger/Edition.

Nigl, Franz; Li, Shuguang; Blum, Jeremy E.; Lipson, Hod (2013) "Structure-reconfiguring robots". *IEEE Robotics & Automation Magazine*, September.

Nihon Biosh Corporation http://www.bisoh.co.jp/en/product/mainten-ance/index.html (Accessed 21 January 2014).

Nishiyama, K.; Kobayashi, K. (2006) "Development of automatic chimney cleaning and brick dismantling equipment". In *Proceedings of the 23rd International Symposium on Automation and Robotics in Construction*, pp. 816–821. Tokyo: International Association for Automation and Robotics in Construction.

nLink http://www.nlink.no/ (Accessed 31 December 2015).

Obayashi Company Brochure "Autoclaw and autoclamp for steel beam and column erection". Tokyo.

Obayashi Company Brochure "Automatic laser beam-guided floor robot". Tokyo.

Obayashi Company Brochure "Clean room inspection and monitoring robot – CRIMRO". Tokyo.

Obayashi Company Brochure "Wall inspection robot – KABEDOHA".

Parietti, Federico; Asada, H. Harry (2014) "Supernumerary robotic limbs for aircraft fuselage assembly: Body stabilization and guidance by bracing". In *2014 IEEE International Conference on Robotics and Automation (ICRA)*, pp. 1176–1183. Hong Kong.

PERI GmbH http://www.peri.de/ (Accessed 31 December 2015).

Phoenix Aerial Systems, Inc. http://www.phoenix-aerial.com/ (Accessed 31 December 2015).

Pritschow, G.; Dalacker, M; Kurz, J.; Gaenssle, M. (1995) "Technological aspects in the development of a mobile bricklaying robot". In *Proceedings of the 12th International Conference on Automation and Robotics in Construction*, pp. 281–290. Warsaw, Poland: International Association for Automation and Robotics in Construction.

Pritschow, G.; Kurz, J.; Fessele, T.; Scheurer, F. (1998) "Robotic on-site construction of mansonry". In *Proceedings of the 15th International Symposium on Automation and Robotics in Construction*, pp. 55–64. Munich, Germany: International Association for Automation and Robotics in Construction.
RIEGL http://www.riegl.com/ (Accessed 31 December 2015).
Robosoft, S. A. (2013) "Glass roof cleaning". http://www.robosoft.com/robotic-solutions/cleanliness/glass-roof.html (Accessed 1 January 2016).
Robotics & Intelligent Construction Automation Lab (RICAL), Georgia Institute of Technology http://rical.ce.gatech.edu/ (Accessed 31 December 2015).
ROBOTSYSTEM http://www.robotsystem.cz/ (Accessed 31 December 2015).
Robotus http://www.robotus.lt/technology-department/robotic-power-trowel-15/en/ (Accessed 31 December 2015).
Roca, D.; Lagüela, S.; Díaz-Vilariño, L.; Armesto, J; Arias, P. (2013) Low-cost aerial unit for outdoor inspection of building façades. *Automation in Construction*, 36: 128–135.
Saidi, K. S.; Bunch, R.; Lytle, A. M.; Proctor, F. (2006) "Development of a real-time control system architecture for automated steel construction". In *Proceedings of the 23rd International Symposium on Automation and Robotics in Construction*, pp. 412–417. Tokyo: International Association for Automation and Robotics in Construction.
Sandvik Mining and Construction http://www.miningandconstruction.sandvik.com/de (Accessed 31 December 2015).
Yun, Seungkook; Hjelle, David Alan; Schweikardt, Eric; Lipson, Hod; Rus, Daniela (2009) "Planning the reconfiguration of grounded truss structures with truss climbing robots that carry truss elements". *In Proceedings of IEEE International Conference on Robotics and Automation, 2009 (ICRA09)*, pp. 1327–1333. Kobe, Japan: IEEE.
Sherman, P. (1988) "Japanese construction R&D: Entree into US market". *Journal of Construction Engineering and Management*, 114.1: 133–143.
Shimizu Company Brochure "Ceiling panel positioning robot". Tokyo: Shimizu.
Shimizu Company Brochure "Civil works robots". Tokyo: Shimizu.
Shimizu Company Brochure "Radio control auto-release clamp Mighty Shackle ACE". Tokyo: Shimizu.
Shimizu Company Brochure. "Remote assembly manipulator for steel beams – Mighty Jack". Tokyo: Shimizu.
Shimizu Company Brochure "Steel beam/column assembly robot". Tokyo: Shimizu.
Shimizu Technology Division Brochure "Automatic spray System OSR1". Tokyo: Shimizu Corporation. Tokyo: Shimizu.
Silver, M. (2014) "Off-road fabrication: Architecture, peripatetic mobility and machine vision". In F. Gramazio; M. Kohler; S. Langenberg (eds.), *Fabricate 2104*. Zürich: GTA Verlag.
Skaff, S.; Staritz, P.; Whittaker, W. (2001) "Skyworker: Robotics for space assembly, inspection and maintenance". In *Proceedings of Space Studies Institute Conference*. Mojave, CA.
Slocum, A. H.; Demsetz, L.; Levy, D.; Schena, B.; Ziegler, A. (1987) "Construction automation research at the Massachusetts Institute of Technology". In *Proceedings of the 4th Internationlal Symposium on Automation and Robotics in Construction (ISARC)*, pp. 222–244. Haifa, Israel: International Association for Automation and Robotics in Construction.
Takenaka Company Brochure "Robots-automation of construction". Tokyo: Takenaka.
Takeno, M.; Matsumura, A.; Sakai, Y. (1989) "Practical uses of painting robot for exterior walls of high rise buildings". In *Proceedings of the 6th International Symposium on Automation and Robotics in Construction*, pp. 285–292. San Francisco: International Association for Automation and Robotics in Construction.
Tanaka, K.; Sakamoto, S.; Abe, Y (2006) "Development of an air volume measuring instrument: WINSPEC". In *Proceedings of the 23rd International Symposium on Automation and Robotics in Construction*, pp. 210–214. Tokyo: International Association for Automation and Robotics in Construction.
Tani, Y.; Nakano, M.; Okoshi, M.; Maeda, S.; Isa; H. (1996) "Research and development of automatic system for open caisson method". In *Proceedings of the 13th International Symposium*

on Automation and Robotics in Construction, pp. 323–332. Tokyo: International Association for Automation and Robotics in Construction.

Taylor, M.; Wamuziri S.; Smith, I. (2003) "Automated construction in Japan". In *Proceedings of ICE 2003*, pp. 34–41. Japan

Teizer, J.; Siebert, S. (2014) "Mobile 3D mapping for surveying earthwork projects using an unmanned aerial vehicle (UAV) system". *Automation in Construction*, 41: 1–14.

Terada, Y.; Murata, S. (2005) "Automatic modular assembly system". *The International Journal of Robotics Research*, 27(3–4): 445–462.

Terada, Y.; Murata, S. (2008) "Automatic modular assembly system and its distributed control". *The International Journal of Robotics Research*, 27(3–4): 445–462. SAGE Publishing: March/April 2008, doi: 10.1177/0278364907085562

Terauchi, S.; Miyajima, T.; Miyamoto, T.; Arai, K.; Takizawa, S. (1993) "Development of an exterior wall painting robot capable of painting walls with indentations and protrusions". In *Proceedings of the 10th International Symposium on Automation and Robotics in Construction*, pp. 285–292. Houston, TX: International Association for Automation and Robotics in Construction.

Ueno, T. (1998) "Automation and robotics in construction in Japan – State of the art". In *Proceedings of the 15th International Symposium on Automation and Robotics in Construction*, pp. 33–36. Munich: International Association for Automation and Robotics in Construction.

Ueno, T.; Kajioka, Y.; Sato, H.; Maeda, J.; Okuyama, N. (1988) "Research and development of robotic systems for assembly and finishing work". In *Proceedings of the 5th International Symposium on Automation and Robotics in Construction*, pp. 279–288. Tokyo: International Association for Automation and Robotics in Construction.

Vanku B. V./ Tiger-Stone http://tiger-stone.nl/ (Accessed 31 December 2015).

Volvo CE – Volvo Construction Equipment http://www.volvoce.com/ (Accessed 31 December 2015).

Vuzix (2015) Eyewear technology. http://www.vuzix.com/consumer/products_m2000ar/ (Accessed 17 May 2015).

Wachsmann, K. (1969) *Turning Point of Building*. New York: Van Nostrand Reinhold.

Wang, C.; Cho, Y. K.; Kim, C. (2015) Automatic BIM component extraction from point clouds of existing buildings for sustainability applications. *Automation in Construction*, 56: 1–13.

Werfel, J. (2012) "Collective construction with robot swarms". Boston: Wyss Institute for Biologically Inspired Engineering, Harvard University.

Werfel, Justin; Petersen, Kirstin; Nagpal, Radhika (2014) "Designing collective behavior in a termite-inspired robot construction team". *Science*, 343(6172): 754–758.

Wickström, G.; Niskanen, T.; Riihimäki, H. (1985) "Strain on the back in concrete reinforcement work". *British Journal of Industrial Medicine*, 42(4): 233–239.

Xiong, X.; Adan, A.; Akinci, B.; Huber, D. (2013) "Automatic creation of semantically rich 3D building models from laser scanner data". *Automation in Construction*, 31: 325–337.

Yang, B; Chen, C. (2015) "Automatic registration of UAV-borne sequent images and LiDAR data". *ISPRS Journal of Photogrammetry and Remote Sensing*, 101: 262–274.

Yi, K. Y.; Lee, J. E.; Chung, S. Y.; Han, S. S. (2004) "Development of a revolutionary advanced auto-shackle". In *Proceedings of International Symposium on Automation and Robotics in Construction 2004*. Jeju, South Korea: International Association for Automation and Robotics in Construction.

Yim, M., Roufas, K.; Duff, D., Zhang, Y., Eldershaw, C., Homans, S. (2003) "Modular reconfigurable robots in space applications". *Autonomous Robots*, 14: 225–237.

Yoshida, N.; Kanagawa, T.; Tani, Y.; Oda, Y. (1997) "Development of an automatic-oriented sheltered building construction system". In *Proceedings of the 14th International Symposium on Automation and Robotics in Construction*, pp. 129–138. Pittsburgh, PA: International Association for Automation and Robotics in Construction.

Yoshida, T.; Ueno, H. (1990) "Development of self-mobile space manipulator system". In *Proceedings of the International Symposium on Artificial Intelligence, Robotics and Automation in Space*. Kobe, Japan.

Yoshida, T.; Ueno, T.; Nonaka, M.; Yamazaki, S. (1984) "Development of spray robot for fireproof cover work". In *Proceedings of the 1st International Symposium on Automation and Robotics in Construction*. Pittsburgh, PA: International Association for Automation and Robotics in Construction.

Zhang, J.; Khoshnevis, B. (2012) "Optimal machine operation planning for construction by contour crafting". *Automation in Construction*, 29: 50–67.

Zykov, V.; Mytilinaios, E.; Adams, B.; Lipson, H. (2005) "Self-reproducing machines", *Nature*, 435(7038): 163–164.

Zykov, V.; Mytilinaios, E.; Desnoyer,, M.; Lipson, H. (2007) "Evolved and designed self-reproducing modular robotics". *IEEE Transactions on Robotics*, 23(2): 308–319.

Zykov, V.; Phelps, W.; Lassabe, N.; Lipson, H. (2008) "Molecubes extended: Diversifying capabilities of open-source modular robotics". In IROS-2008 Self-Reconfigurable Robotics Workshop. Nice.

Glossary

Alignment and accuracy measurement system (AAMS): An AAMS creates a feedback loop between a system that measures how accurately components are positioned and an alignment system (e.g., a motorized unit attached temporarily to the joint of a column component) that automatically moves or aligns the *component* into the desired final position.

Assembly: In this book, assembly refers to the production of higher-level components or final products out of *parts* and lower-level *components*. The process of assembly of an individual *part* or *component* to a larger system involves positioning, alignment, and fixation operations. *Upstream* processes dealing with the generation of elements for assembly are referred to as *production*.

Automated guided vehicle (AGV): Computer-controlled automatic or robotic mobile transport or logistics vehicle.

Automated/robotic on-site factory: *Structured environment* (factory or factory-like) setup at the place of *construction*, allowing *production* and *assembly* operations to be executed in a highly systemized manner by, or through, the use of machines, automation, and robot technology.

Batch size: The amount of identical or similar products produced without interruption before the manufacturing system is substantially changed to produce another product. Generally speaking, low batch sizes are related to high fixed costs and high batch sizes are related to low fixed costs.

Building component manufacturing (BCM): BCM refers to the transformation of *parts* and low-level *components* into higher-level *components* by highly mechanized, automated, or robot-supported industrial settings.

Building integrated manufacturing technology: Automation technology, microsystem technology, sensor systems, or robot technology can be directly integrated into buildings, units, or components as a permanent system. Technology used to manufacture the building can thus become a part of the building technology.

Bundesministerium für Bildung und Forschung (BMBF): Federal Ministry of Education and Research in Germany. The BMBF funds education, research, and technical development in a multitude of industrial fields.

Capital intensity: The capital intensity (also referred to as workplace cost) is calculated by dividing the capital stock (assets, devices, and equipment used to transform/manufacture the outcome) by the number of employees in the industry.

Chain-like organization: In a chain-like organization, the *flow of material* between individual workstations is highly organized and fixed, and a material transport system linking the stations exists.

Climbing system (CS): Automated/robotic on-site factories require, especially in the manufacturing of vertically oriented buildings, a system that allows the *sky factory* (SF) to rise to the next floor, once a floor level has been completed. Most SFs, therefore, rest on stilts that transfer the loads of the SF to the building's bearing structure, or to the ground. Other CSs are able to climb along a central core, pushing up the building or, in the case of the manufacture of horizontally orientated buildings, for example, to enable the factory to move horizontally. In some cases, in addition to climbing, CSs are used to provide a fixture or template for the positioning of components by *manipulators*. Because of the enormous forces necessary to lift SFs, hydraulic systems and screw presses are common actuation systems.

Closed loop resource circulation: Systems for avoiding waste and reduction of resource consumption, by integrating concepts such as reverse logistics, remanufacturing, and recycling. Material or product that flows on a factory, utilization, and deconstruction level can be related back to the manufacturing system to close the loop.

Closed sky factory (CSF): *Sky factory* that completely covers and protects the workspace in *automated/robotic on-site factories*, thus allowing the installation of a fully *structured environment* that erases the influence of parameters that cannot be 100% specified (e.g., rain, wind).

Component: In this book, in a hierarchical modular structure, components can be divided into lower-level components and higher-level components. Components consist of subelements of *parts* and lower-level components. Higher-level components can be assembled into *modules* and *units*.

Component carriers: Component carriers and pallets (special types of component carriers) play an important role in logistics. In many cases, parts, components, or final products cannot be directly handled or manipulated by the logistics system. Component carriers and pallets act as mediators between the handled material and the actual logistics system.

Computer-aided design/computer-aided manufacturing (CAD/CAM): From the 1980s on, the novel and highly interdisciplinary research and application field CAD/CAM was formed, which aimed at integrating computerized tools and systems from the planning and engineering field with manufacturing and machine control systems to allow for a more-or-less direct use of the digital design data for automated and flexible manufacturing. The field evolved further towards *computer-integrated manufacturing*.

Computer-aided quality management (CAQM): Control of quality by software, made possible through the linking of manufacturing systems with computer systems.

Computer-integrated manufacturing (CIM): From the 1990s on, the combined *computer-aided design/computer-aided manufacturing* approach evolved into the CIM approach. The focus was then broader and the idea was that more and more fields and tools, and also business economic issues (e.g., computer-aided forecasting or demand planning), could be integrated by computerized systems to form continuous process and information chains in manufacturing that span all value-adding nodes in the value system.

Connector system: The development of connector systems that connect complex *components* in a robust way to each other is a key element in complex products such as cars, aircraft, and buildings in particular. To support efficient *assembly*, connector systems can, for example, be compliant or plug-and-play–like. Connector systems can also be designed to support efficient disassembly, remanufacturing, or recycling.

Construction: Activities necessary to build a building on-site. Construction, in this book, is interpreted as being a manufacturing process, and accordingly buildings are seen as "products".

Cycle time: Important on the workstation level: The cycle time refers to the time allowed for all value-adding activities performed by humans and machines at a workstation within a network of workstations.

Degree of freedom (DOF): In a serial kinematic system each joint gives the systems, in terms of motion, a DOF. At the same time, the type of joint restricts the motion to a *rotation* around a defined axis or a *translation* along a defined axis.

Depth of added value: The depth of added value (e.g., measured as a percentage of the total cost of the product) refers to the total amount of value-adding activities, and thus in general to the amount of value-adding steps, realized by the *original equipment manufacturer* (OEM) or final integrator. A high depth of added value means that a large number of value-adding activities are being realized by the OEM (e.g., Henry Ford). A low depth of added value means that a low number of value-adding activities is being realized by the OEM (e.g., Dell, Smart).

Design for X (DfX): DfX strategies aim at influencing design-relevant parameters to support production, assembly maintenance, disassembly, recycling, and many other aspects related to a product's life cycle. In this book, DfX strategies are classified into four categories: DfX related to *production/assembly*, to product function, to product end-of-life issues, and to business models. In this book, *robot-oriented design* is seen as an augmentation or extension of conventional DfX strategies, consequently aiming more at the efficient use of automation and robotic technology in all four categories.

Deutscher Kraftfahrzeug-Überwachungs-Verein (DEKRA): Major German consultant and surveyor association that evaluates technical artefacts, such as cars and buildings, and defines quality and the causes of defects.

Efficiency: Efficiency can be defined as the relationship between an achieved result and the combination of factors of production. Whereas *productivity* expresses an input-to-output ratio, with a focus on a single factor of production, efficiency considers multiple factors and their combination and interrelation. Productivity can be an indication of efficiency, and efficiency itself for economic feasibility.

End-effector: The element of machines, automation systems, or robot technology that makes contact with the object to be manipulated in *manufacturing* is called the end-effector. In most cases end-effectors are modularly separable from the base system. End-effectors have a certain degree of *inbuilt flexibility*.

Factory external logistics (FEL): FEL refers to logistics systems that connect the supply network to the factory integrating and assembling the supplied *parts*, *components*, *modules*, or *units*. FEL influences the organization of the manufacturing system, *factory internal logistics*, and the factory layout.

Factory internal logistics (FIL): FIL refers to logistics systems that manipulate *parts*, *components*, *modules*, *units*, or the finished product within a manufacturing setup or factory, for example, for the transportation between various stations. Other examples include mobile and non-rail–guided transport systems, overhead crane-type material transportation systems, fixed conveyor systems allowing a component carrier or the product itself to travel in a horizontal direction in fixed lanes, and fixed conveyor systems allowing a component carrier or the product itself to travel in a vertical direction in fixed lanes. Novel cellular logistics robots combine capabilities of unrestricted mobility with horizontal and vertical transport capabilities and can travel freely and self-organize with other systems.

Factory roof structure: Structure that allows the workspace on the construction site to be covered (and therefore to be protected from outside influences such as wind, rain, or sun) and thus creates the basis for a *structured environment*. Often used as a platform for the attachment of other subsystems, such as a *climbing system*, *horizontal delivery system*, and *overhead manipulators*.

Final integrator: In this book, a final integrator refers to an entity in a value chain or value system that integrates major components into the final product. Within the OEM model, the final integrator is called *original equipment manufacturer*.

Fixed-site manufacturing: *Off-site manufacturing* or *on-site manufacturing* system that stays at a fixed place during final *assembly*.

Floor erection cycle (FEC): Time necessary to erect and finish (including technical installations and general interior finishing) a standard floor with an *automated/robotic on-site factory*.

Flow-line organization: In a flow-line organization, individual workstations do not have a fixed *flow of material*, but a general directional *flow of material* (e.g., within a factory segment or a factory).

Flow of material: Refers to material and product streams in relation to space and time that take place during the completion of a specific product in a manufacturing system and the supply network connected to it. The *efficiency* of the flow of material is determined by the arrangement of equipment, the factory layout, and the logistics processes.

Frame and infill (F&I): F&I strategies are used in a variety of industries, including the aircraft, automotive, and building industries. The idea of an F&I strategy is to use a bearing frame structure as a base element that is subsequently equipped with *parts*, *components*, systems, *modules*, and so forth during the manufacturing process. The frame thus functions as a *component carrier*. In the aircraft industry, the fuselage is interpreted as such a frame; in the automotive industry it is the car body or chassis; and in the building industry it can be seen, for example, in the form of two-dimensional (e.g., Sekisui House) or three-dimensional steel frames (Sekisui Heim).

Ground factory (GF): *Structured environment* (factory or factory-like) setup on the construction site on the ground level of the building as part of an *automated/robotic on-site factory*.

Group-like organization: In a group-like organization, individual workstations are bound together in groups. Those groups can refer to workstations with similar means of production or to workstations with complementary *means of production*. The *flow of material* between those groups can be either fixed or flexible.

Horizontal delivery system (HDS): System that transports, positions, and/or assembles *parts/components* on the construction site on a floor level.

Idle time: The unproductive standstill of a machine from end of completion to the beginning of the processing of the next material. Bottleneck operations, for example, may – when workstations are directly connected without a buffer – lead to material having to wait for a certain time until the next material can be processed and to an unproductive standstill of other workstations that are faster in processing the material.

Inbuilt flexibility: The changes in a manufacturing system can be realized without major physical or modularization enabled changes (e.g., exchange of systems, workstations, robots, *end-effectors*), but by reprogramming the existing system instead. A standard robot with 6 *degrees of freedom* (6-DOF robot) with an end-effector for welding, for example, has a high degree of flexibility and can be reprogrammed for a huge variety of welding operations within a given workspace.

Joint of a manipulator: A *manipulator* consists of at least one kinematic pair consisting of two rigid bodies (links) interconnected with a joint. The following types of joints can be distinguished: revolute joint, prismatic joint, and spherical joint.

Just in sequence (JIS): Various *parts*, *components*, and *products* are delivered from *upstream* to *downstream* workstations in the sequence in which they are handled or processed when they reach the *downstream* work stations. JIS can be performed internally within a factory or in relation to a supplier of an *original equipment manufacturer* (OEM). JIS is in most cases closely connected to *just in time* (JIT).

Just in time (JIT): Stocks and buffers are eliminated, and *parts*, *components*, and products are delivered from *upstream* to *downstream* workstations at the right time and at the right quantity. JIT can be performed internally within a factory or in relation to a supplier of an *original equipment manufacturer*. JIT is in most cases closely connected to *just in sequence*.

Kinematic base body: The combination of links and joints forms kinematic bodies that allow basic manipulation operations within a geometrically definable work space (e.g., Cartesian *manipulator*, gantry *manipulator*, cylindrical *manipulator*, spherical *manipulator*). Those kinematic base bodies consider mainly the first three axes, and thus refer mainly to *positioning* activity. For *orientation*, further *degrees of freedom* and kinematic combinations can be added on top of those base bodies.

Kinematics: Kinematics focuses on the study of geometry and motion of automated and robotic systems. It describes parameters such as position, velocity, and acceleration of joints, links, and *tool centre points* to generate mathematical models creating the basis for controlling the actuators and for finding optimized trajectories for the motions of the system. *Manipulators* are a kinematic system consisting of a multitude of kinematic subsystems, of which the kinematic pair is the most basic entity.

Large-scale prefabrication (LSP): *Off-site manufacturing* of high-level *components*, *modules*, or *units* in very large quantity by a production-line–based, automation and robotics-driven factory or factory network, interconnected in an *OEM-like integration structure*.

Link of a manipulator: A *manipulator* consists of at least one kinematic pair, consisting of two rigid bodies (links) interconnected with a *joint*.

Logistics systems: Logistics can be defined as the transport of material within manufacturing systems and supply networks. Logistics is a kind of manipulation of an object by humans, tools, machines, automation systems, and robots (or combinations of those), positioning and orientating objects to be transported or processed in a three-dimensional space. However, logistics operations do not change or transform the material directly. Logistics systems can be characterized according to various scales, such as assembly system scale, factory internal scale (*factory internal logistics*), factory external scale (*factory external logistics*).

Manipulator: In this book series, a manipulator refers to a system of multiple links and joints that performs a kinematic motion. Depending on the ratio of autonomy and intelligence, manipulators can be machines, automated systems, or robots.

Manufacturing: In this book, manufacturing refers to systems that produce products. Manufacturing integrates *production* (parts or low-level component production) and *assembly* processes.

Manufacturing lead time: Time necessary to complete a product within a given manufacturing system, factory, or factory network.

Mass customization (MC): MC strategies combine advantages of *workshop-like* and *production-line*–like manufacturing, and thus product differentiation–related competitive advantages, with mass-production–like efficiency. On the product side, MC demands that a product combines customized and standardized elements, for example, through *modularity*, *platform strategies*, and *frame and infill* strategies to be able to efficiently produce it in an industrialized manner. On the manufacturing side, MC demands highly flexible machines, automation systems, or robot technology that removes the need for human labour in the customization process.

Material handling, sorting, and processing yard (MHSPY): Subsystem of an *automated/robotic on-site factory*; often related to the *ground factory*. An MHSPY can be a covered environment and/or can be equipped with *overhead manipulators* and allows the simplification or automation of the picking up of *components* from delivering *factory external logistics* in a *just in time* and *just in sequence* manner. An MHSPY can also be used to transform *parts* and low-level *components* into higher-level components on-site. In *automated/robotic on-site factories* used to deconstruct buildings, MHSPYs can be used to transform higher-level *components* into lower-level *components* and *parts*.

Means of production: Means of production can be classified into human resources, equipment, and material to be transformed.

Modular flexibility: When the change of a product or the variation of a product is so intense that the *inbuilt flexibility* of a manufacturing system, a machine, or an *end-effector* cannot cope with it, a rearrangement or extension of the manufacturing system on the basis of *modularity* becomes necessary. *Modularity* can be generic (predefined process or system modules) or unforeseen (use/design of completely new modules, new configurations).

Modularity: Modularity refers to the decomposition of a structure or system into rather independent subentities. It can cover the functional realm as well as the physical realm. If structures or systems are nearly impossible to decompose, on both functional and physical levels, the artefacts are referred to as "integral". If systems can clearly be decomposed on both functional and physical levels, artefacts are referred to as "modular". Clear modularity is, in construction practice, still a rare phenomenon, and conventional buildings show basic characteristics of integral product structures. *Automated/robotic on-site factories*, however, require strict modularity.

Module: In this book, in a hierarchical modular structure, modules represent elements on a hierarchical level above high-level *components*. *Parts*, *components*, and modules can be assembled into *units*, which are ranked higher than modules.

nth, $n-1$, $n-2$, $n-X$ floors: Inside the main factory (e.g., a *sky factory*) of *automated/robotic on-site factories*, work (component installation, welding, interior finishing, etc.) is done in parallel on several floors (n-floors). The nth floor represents the uppermost floor in which work takes place in parallel and the $n-X$ floors represent the floors below this floor in which work takes place in parallel.

OEM-like integration structure: Value systems or *parts/components* integration structures that do not fully follow the *OEM model* but show characteristics of it.

OEM model: An *original equipment manufacturer* relies on suppliers, which, according to their rank in the supply chain, are called Tier-n suppliers. The model explains the general *flow of material* as well as the flow of information during development of the product and its subcomponents.

Off-site manufacturing (OFM): *Components* or complete products are manufactured in a *structured environment* distant from the final location where they are finally used. *Components* or complete products can be packed and shipped or are mobile (e.g., car, aircraft).

One-piece-flow (OPF): OPF refers to a highly systemized and *production-line–* based manufacturing system in which each *component* or product assembled can be different.

On-site manufacturing (ONM): Products such as buildings, towers, and bridges have to be produced on site by ONM systems at the location at which they are to be finally used as they cannot be moved or shipped as an entity.

Open sky factory (OSF): *Sky factory* covering and protecting the workspace in *automated/robotic on-site factories*, and, in contrast to *closed sky factories* allows only the installation of a partly *structured environment* that at least (compared to conventional construction) minimizes the influence of parameters that cannot be 100% specified/foreseen (e.g., rain, wind).

Original equipment manufacturer (OEM): Integrates and assembles *components* and subsystems coming from sub-factories and suppliers to the final product within the *OEM model*. Companies or entities in the value chain that do not fully follow the *OEM model* but show characteristics of it are also referred to as *final integrators*.

Overhead manipulators (OMs): OMs operate within off- or on-site *structured environments* and in *automated/robotic on-site factories* are often the central elements of the *horizontal delivery system*. On the one hand, OMs (e.g., gantry-type OMs) allow the precise manipulation of components of extreme weights and at high speed, which cannot, for example, be accomplished by conventional industrial robots such as anthropomorphic manipulators. On the other hand, OMs require a simplification of the assembly process by *robot-oriented design*, as their workspace and their ability to conduct complex positioning and orientation tasks are limited.

Part: In this book, in a hierarchical modular structure, parts represent elements on a hierarchical level below *components*.

Performance multiplication effect (PME): Once significant productivity increases in an industry can be achieved (i.e., by switching from crafts-based to machine-based manufacturing), an upward spiral starts: high productivity can become a driver of the financing elements for innovations related to even better machines, processes, and products and thus even higher productivity. This phenomenon was/can be observed in many non-construction industries (e.g., textile industry, automotive industry, shipbuilding) and is in this book series referred to as the PME.

Platform strategy: A platform is a basic framework; a set of standards, procedures, or parts; or a basic structure that contains core functions of a product. A platform allows for the highly efficient production of customized products, as it allows for the platform to be mass-produced and to wear individual modules on top of it, which can be customized or personalized.

Positioning and orientation: For unrestricted positioning of an object within a defined space, or within x, y, and z coordinates, at least 3 *degrees of freedom* are necessary (also referred to as forward/back, left/right, up/down). For unrestricted orientation of an object around a *tool centre point*, at least 3 *degrees of freedom* are necessary (also referred to as yaw, pitch, and roll).

Production: In this book, production refers to the generation of basic *parts* or low-level *components*. It includes transformation of raw material into *parts*. *Downstream* processes dealing with the joining of elements generated within production are referred to as *assembly*. *Manufacturing* includes production and *assembly* processes.

Production line–like organization: In a production line–like organization, the flow of material between individual workstations is fixed; a material transport system links the stations and the *cycle times* of the individual workstations are synchronized.

Productivity: Productivity = Output (quantity)/Input (quantity). Productivity quantitatively expresses an input-to-output ratio, with a focus on a single (input) means of production or a single (input) factor of production. Productivity indices concerning the type of factor are, for example, work, capital, material, resource, and machine productivity.

Pulling production: Refers to a production system in which products are manufactured only on the basis of actual demand or orders. *Parts*, *components*, and products required are pulled from *upstream*, according to the actual demand. It might refer to the whole manufacturing system, as well as to individual workstations or groups of workstations. Examples: *Toyota Production System*, Sekisui Heim, Toyota Home.

Pushing production: Refers to the continuous production of elements/products in a certain fixed amount based on predictions or assumptions. Without taking into consideration the actual demand in *downstream* process steps, *parts*, *components*, and products are pushed through individual stations. It might refer to the whole manufacturing system or to individual workstations or groups of workstations. Example: Henry Ford's mass production.

Radio frequency identification tag (RFID): RFID tags are inexpensive tags that can be attached to *components*, *modules*, *units*, or products. RFID readers can be integrated into floors or placed over gates and can then identify the object passing by. They can be distinguished between simple low-cost passive tags and more complex active tags. Advanced readers can read multiple tags at once.

Real-time economy (RTE): Macroeconomic view of the impact of the multitude of changes our economy, manufacturing technology, and the relation between customers and businesses undergo. It targets the fulfilment of customer demands and requests in near real time. Products and services are processed within a few hours and delivered within a few days.

Real-time monitoring and management system (RTMMS): Data from sensor systems, as well as from the servomotors/encoders of the *vertical delivery system* (VDS) and *horizontal delivery system* (HDS), along with information obtained from cameras monitoring all activities (including human activities) in *automated/robotic on-site factories* are used to create a real-time representation of equipment activity and of construction progress. Furthermore, barcode systems often allow the representation and optimization of the *flow of material*, allowing equipment (such as VDS, HDS, or OM) to identify the *component* being processed. In most cases, real-time monitoring and management is done in a fully computerized on-site control centre. An RTMMS simplifies progress and quality control and reduces management complexity.

Re-customization: Remanufacturing strategy that allows a building to be disassembled and for major *components* or *units* to be refurbished and equipped with new *parts* or *modules* on the basis of mechanized or automated manufacturing systems to meet changed or new (individual) customer demands.

Robot-oriented design (ROD): ROD is concerned with the co-adaptation of construction products and automated or robotic technology, so that the use of such technology becomes applicable, simpler, or more efficient. The concept of ROD was first introduced in 1988 in Japan by T. Bock and served later as the basis for automated construction and other robot-based applications.

Rotation: A term used to describe a kinematic structure. A revolute joint allows an element of a machine or manipulator to rotate around an axis and in a serial kinematic system adds 1 *degree of freedom* to the system.

Selective compliance articulated robot arm (SCARA): Developed by Yamanashi University in Japan in the 1970s. It combines two revolute and one prismatic joint so that all motion axes are parallel. This configuration and the thus enabled allocation of the actuators are advantageous for the stiffness, repeatability, and speed with which the robot can work. Owing to its simplicity, the SCARA is also a relatively cheap robot system. It laid the foundation for the efficient and cheap production, and thus the success of, for example, Sony's Walkman.

Single-task construction robot (STCR): STCRs are systems that support workers in executing one specific construction process or task (such as digging, concrete levelling, concrete smoothening, and painting) or take over the physical activity of human workers that would be necessary to perform one process or task.

Sky factory (SF): Structured environment (factory or factory-like) setup on the construction site as part of *automated/robotic on-site factories*. SFs cover the area where building *parts* and *components* are joined to the final product and rise vertically with the upper floor of a building through a *climbing system*. SFs can enclose and protect the work environment completely (*closed sky factory*) or only partly (*open sky factory*).

Slip forming technology: Moving or self-moving form that allows casting concrete structures such as columns, walls, or towers on site in a systemized manner on the construction site.

Stilts: In this book, stilts refer to elements of automated/robotic on-site factories. The *sky factories* of *automated/robotic on-site factories* often use stilts (made of steel) integrated within the *climbing system* to be rested on the building that they are manufacturing. Stilts can be lifted and lowered via the *climbing system*, thus allowing the *sky factory* to move on top of the building's steel column structure.

Structured environment (SE): In factories or factory-like environments, work tasks, workspaces, assembly directions, and many other parameters (e.g., climate, light, temperature) can be standardized and precisely controlled. The structuring of an environment creates the basis for the efficient use of machines, automation, and robot technology. The structuring of an environment includes the protection from uncontrollable factors such as wind, rain, sun, and non-standardized human work activity.

Superstructure: The concept of dividing a building into superstructure and substructure is an approach that introduces the concept of hierarchies to a building's structure and *components* and thus can serve as a basis for possible *modularity* in construction. Goldsmith introduced the idea of making the transmission of forces within high-rise buildings by the superstructure to a clearly visible architectural element in his thesis (1953). A superstructure can serve as a platform or frame that allows customization by further infill and is thus closely connected to *frame and infill* strategies.

Supply chain: The supply chain connects value-added steps and transformational processes across the border of individual factories or companies. Its aim is to interconnect all processes and workstations to complete a product informationally and physically, in order to create uninterrupted on-demand or in-stock *flow of material*.

Sustainability in manufacturing: Manufacturing systems can be designed to be efficient and to equally meet economic, environmental, and social demands. In this book, sustainability in manufacturing refers predominantly to the ability of a manufacturing system to reduce consumption of resources and the generation of waste.

Technology diffusion: Technology diffusion describes the step-by-step spread of a technology throughout industry or as, for example, computer technology throughout society. To simplify the adoption of novel technologies and increase their application scope over time, novel technologies have to be made less expensive and less complex and be split into individual modular elements. Technology diffusion therefore often is accompanied by a switch from centralized to rather decentralized applications of the novel technology.

Tier-*n* supplier: Suppliers are, according to their rank in the supply chain, called Tier-*n* suppliers. A Tier-1 supplier is a first-rank supplier that relies on *components* from a Tier-2 supplier; a Tier-2 supplier relies on *components* from Tier-3 suppliers; and so on.

Tool centre point (TCP): The *end-effector* is a tool that is carried by the kinematic system. For each end-effector, a tool TCP is defined as the reference point for kinematic calculations.

Toyota Production System (TPS): The TPS is a logical and consequent advancement of the concept of mass production to a more flexible and adaptive form of demand-oriented manufacturing, developed by Toyota between the 1960s and 1970s. From the 1980s, TPS principles gained worldwide recognition and today they form the conceptual basis for manufacturing systems around the world. Concepts such as *just in time*, Kaizen, *Kanban*, *pulling production*, failure-free production, and *one-piece flow* have their origins in the TPS.

Transformational process: Any organization and its manufacturing system transform inputs (information, material) into outputs (products, services). The transformation is performed by the organization's structural setting and its means of production, resulting in a specific combination and interaction of workers, machines, material, and information.

Translation: A term used to describe a kinematic structure. A prismatic joint allows an element of a machine or manipulator to move in a given trajectory along an axis and, in a serial kinematic system, adds 1 *degree of freedom* to the system.

Tunnel boring machine (TBM): TBMs mechanize and automate repetitive processes in tunnel construction. TBMs are self-moving underground factories that more or less automatically perform excavation, removal of excavated material, and supply and positioning of precast concrete segments. TBMs are equal in many ways to automated construction sites (and the production of a building's main structure by *automated/robotic on-site factories*).

24/7-mode: The operation of a factory, processes, or equipment without major interruption 24 hours a day, 7 days a week. Requires the work environment to be structured (*structured environments*) and protected from influencing factors such as weather and the day/night switchover.

Unit: In this book, units refer to high-level building blocks. *Parts*, *components*, and *modules* can be assembled into units, which are higher ranked than *modules*. Units are completely finished and large three-dimensional building sections, manufactured off site.

Unit method: Sekisui Heim, Toyota Home, and Misawa Homes (Hybrid) break down a building into three-dimensional *units*. Those *units* are realized on the basis of a three-dimensional steel space frame, which, on the one hand is the bearing (steel) structure of the building, and on the other hand can be placed on a production line where it can be almost fully equipped with technical installations, finishing, kitchens, bathrooms (plumbing *units*), and appliances.

Upstream/downstream: Refer in this book to processes or activities in a value chain or manufacturing chain that are conducted before (upstream) or after (downstream) a certain point.

Urban mining: Refers to strategies that allow the city and especially its building stock to be a "mine" for resources, *parts*, and *components*. Systemized deconstruction of buildings under controlled and structured conditions, as in *automated/robotic on-site factories*, are enablers of urban mining.

Vertical delivery system (VDS): System that transports *parts/components* on the construction site from the ground (e.g., material handling yard) to the floor level where the components are to be assembled. VDSs play an important role in most *automated/robotic on-site factories*.

Workbench-like organization: The product stays at a fixed station in the factory where it is produced or assembled manually or automatically through the use of various tools. The means of production are organized around this one station.

Workshop-like organization: In a workshop-like organization, the product and/or its components flow between workstations. The sequence is not fixed and the times products stay at a certain workstation vary with the product.

"Zero" waste factory: A factory that minimizes resource input and waste output and allows (almost) all generated waste to be recycled.

Index